Study Guide

Kenneth G. Pinzke

Foundations *of* Earth Science

Second Edition

Lutgens & Tarbuck

Illustrated by Dennis Tasa

Prentice Hall

PRENTICE HALL, Upper Saddle River, NJ 07458

Supplement Editor: Wendy Rivers
Special Projects Manager: Barbara A. Murray
Production Editor: Ann Marie Kalajian
Supplement Cover Manager: Paul Gourhan
Supplement Cover Designer: Liz Nemeth
Manufacturing Buyer: Ben Smith

Upper Saddle River, NJ 07458

Printed in the United States of America

10 9 8 7 6 5 4 3 2

ISBN 0-13-010063-3

Prentice-Hall International (UK) Limited, *London*
Prentice-Hall of Australia Pty. Limited, *Sydney*
Prentice-Hall Canada, Inc., *London*
Prentice-Hall Hispanoamericana, S.A., *Mexico*
Prentice-Hall of India Private Limited, *New Delhi*
Prentice-Hall of Japan, Inc., *Tokyo*
Pearson Education Asia, *Singapore*
Editora Prentice-Hall do Brasil, Ltda., *Rio de Janeiro*

CONTENTS

Preface

Keys to Successful Learning

Learning involves much more than reading a textbook and attending classes. It is a process that requires continuous preparation, practice, and review in a dedicated attempt to understand a topic through study, instruction, and experience. Furthermore, most learning experts agree that successful learning is based upon association, repetition, and visualization.

Students often ask questions such as "What is the best way to learn Earth science?" or, "How should I study?" Although there are no set answers to these questions, many successful students use similar study techniques and utilize many of the same study habits. For you to become a successful learner, you need to develop certain skills that include, among others:

1. **Carefully observing the world around you**. Pay attention to the relations between natural phenomena. For example, when studying weather, you might observe and note the direction that storms move in your area, or the relation between air temperature and wind direction. Or when studying minerals and rocks, you may want to list, locate, and identify those that are found in your area.
2. **Being curious and inquisitive**. Ask questions of your instructor, your classmates, and yourself, especially about concepts or ideas you do not fully understand.
3. **Developing critical thinking skills**. Carefully analyze all available information before forming a conclusion and be prepared to modify your conclusions when necessary on the basis of new evidence.
4. **Visualizing**. Take time to create mental images of what you are studying, especially those concepts that involve change through time. If you can "picture" it, you probably understand it.
5. **Quantifying**. Accurate measurement is the foundation of scientific inquiry. To fully comprehend scientific literature you must be familiar with the common systems and units of measurement, in particular the metric system. If you are not already comfortable with the metric system, take time to study and review it. Practice "thinking metric" by estimating everyday measurements such as the distance to the store or the volume of water in a can or pail using metric units.
6. **Being persistent**. Be prepared to repeat experiments, observations, and questions until you understand and are comfortable with the concepts in your own mind.

Remember that an important component of learning is identifying what you *should* know and understand, and what you *do* know and understand about the material presented in the textbook and during class. With this in mind, the following steps may help you become an active, successful learner.

1. **Prepare**. Quickly scan the pages of the textbook chapter and accompanying section of this Study Guide. Try to visualize the organization of the chapter in simple, broad concepts. Read and examine the questions presented a the beginning of each textbook chapter, as well as the chapter overview and learning objectives provided for each chapter in the Study Guide. These will help you focus on the most important concepts the authors present in the chapter.
2. **Read the Assignment**. Read each assignment before attending class. Taking written notes as you read can be and effective means of helping you remember major points presented in the chapter. As you read the assignment, you may also find the following suggestions helpful.
   - To help you visualize concepts, look at all the illustrations carefully and read the captions. Also try to develop mental images of the material you are reading.
   - Pace yourself as you read. Often a single chapter contains more subject matter than can be assimilated in one long reading. Take frequent short breaks to help keep your mind fresh.

- If you find a section confusing, stop and read the passage again. Make sure you understand the material before proceeding.
- As you read, write down questions you may want to ask your instructor during classtime.

3. **Attend and Participate**. To get the most from any subject, you should attend all lecture classes, study sessions, review sessions, and (if required) laboratory sessions.
   - Bring any questions concerning the subject matter to these sessions.
   - Communicate with your instructor and the other students in the class.
   - Learn how to become a good listener and how to take useful class notes. Remember that taking a lot of notes may not be as effective as taking the right notes.

- Come fully prepared for each session.

4. **Review**. Review your class notes and the chapter material by examining the section headings, illustrations, and chapter notes. Furthermore:
   - Examine the review statements presented at the end of the textbook chapter. Then see if you can answer the questions that were presented at the beginning of the textbook chapter. If you can, you have a good comprehension of the chapter material.
   - Indentify the key facts, concepts and principles in the chapter by highlighting or underlining them.
   - Write out the answers to the review questions presented at the end of the textbook chapter.
   - To assess how well you have reviewed, take the short test found at the end of each textbook chapter. For any questions you miss, locate, read, and study the appropriate section in the chapter until you understand the concept.

5. **Studying is one of the most important keys to successful learning**. Very few individuals have the ability to retain vast amounts of detailed information after only one reading. In fact, most of us require seeing and thinking about a concept many times, perhaps over several days, before we fully comprehend it. Consider the following as you study and prepare for exams.
   - As you study, keep "putting the pieces together" by integrating the key concepts and facts as you go along. The chapter outlines presented in the Study Guide may prove useful for reviewing and organizing the major points of each chapter.
   - As you review your textbook notes, lecture notes, and chapter highlights, keep in mind that most introductory science courses emphasize terms and definitions because scientists must have consistent, precise meanings for purposes of communication. To quiz your knowledge of each chapter's key terms, complete the vocabulary review section of the appropriate chapter in the Study Guide. Review your answers often prior to an examination.
   - Complete the comprehensive review section of the appropriate chapter in the Study Guide by writing out complete answers to the questions. By actually writing the answers, you will be able to retain the material with greater detail than simply answering the questions in your mind or copying the answers from the answer key.
   - Several days before an exam, test your overall retention of the chapter material by taking the practice test for the appropriate chapter(s) in the Study Guide. Thoroughly research and review any of the questions you miss.

Using the Study Guide Effectively

This study guide has been written to accompany the textbook *Foundations of Earth Science–Second Edition* by Frederick K. Lutgens and Edward J. Tarbuck. For each chapter of the textbook there is a corresponding Study Guide chapter. For example, Chapter 1 of the Study Guide corresponds to Chapter 1 of the textbook. Each chapter in the Study Guide contains nine parts. They are:

1. *Chapter Overview.* The overview provides a brief summary of the topics presented in the chapter. The overview should be read before reading the textbook chapter.

2. *Learning Objectives.* The learning objectives are a list of what you should know after you have read and studied the textbook chapter and completed the corresponding chapter in the Study Guide. You should examine the objectives before and after you read and study the chapter.

3. *Chapter Review.* The chapter outline presents, in outline form, the major topics of the chapter. It is helpful for both reviewing chapter concepts and preparing for examinations.

4. *Chapter Outline.* The chapter outline presents, in outline form, the major topics of the chapter. It is helpful for both reviewing chapter concepts and preparing for examinations.

5. *Key Terms.* The key terms list includes the new vocabulary terms introduced in the textbook chapter. the page number shown in parentheses ( ) after each term identifies the textbook page where the term first appears.

6. *Vocabulary Review.* This section consists of fill–in–the–blank statements that are intended to test your knowledge of the new terms presented in the textbook chapter. You should make sure you have a good understanding of the terms before completing the remaining sections of the Study Guide chapter. Answers to the vocabulary review are listed in the answer keys located at the back of the Study Guide.

7. *Comprehensive Review.* The comprehensive review contains several questions that test your knowledge of the basic facts and concepts presented in the chapter. In this section you will be asked to write out answers, label diagrams, and list key facts. Answers to the comprehensive review questions are listed in the answer keys located at the back of the Study Guide.

8. *Practice Test.* This section of each Study Guide chapter will test your total knowledge and understanding of the textbook chapter. Each practice test consists of multiple choice, true/false, and written questions. Taking the practice test is one way for you to determine how ready you are for an actual exam. You may find it helpful to write your answers to the practice test questions on a separate sheet of paper and leave the test in the Study Guide unmarked so you can review it several times while preparing for a unit exam, mid–term exam, and/or final exam. Answers to the practice test questions are listed in the answer keys located at the back of the Study Guide.

9. *Answer Key.* Chapter answer keys are located at the back of the Study Guide.

To effectively use the Study Guide, it is recommended that you

- Quickly glance over the appropriate chapter in the Study Guide before reading the corresponding textbook chapter.
- When required, write out complete answers to the Study Guide questions.
- Check your answers with those provided at the back of the Study Guide.
- Research the answers you don't know by looking them up in your textbook and/or class notes. You will learn very little by simply copying the answers from the answer keys.
- Test yourself by taking the practice test provided at the end of each Study Guide chapter.
- Constantly review your answers to all the Study Guide questions.

Earth Science and the World Wide Web

In this era of rapidly expanding information and the global communication network referred to as the Internet, the ability to access, retrieve, store, and use data is a fundamental skill required of Earth scientists. Designed to give you an interactive means to further your studies and effectively utilize the Internet, the authors have prepared a comprehensive *Foundations of Earth Science* companion World Wide Web site that reaches out into cyberspace and brings the myriad of Earth science resources available there closer to you. In addition to providing direct links to numerous Web sites that present interesting and relevant information that supplements and reinforces the topics presented in each chapter, the *Foundations of Earth Science* site is one of the most outstanding educational Internet locations available today. Check out these interactive resources, interesting links, and items for further thought via the Internet at **http://www.prenhall.com/lutgens**.

ACKNOWLEDGEMENTS

My sincere thanks to each of the many individuals who assisted in the preparation of this second edition of the *Study Guide* for *Foundations of Earth Science.* Furthermore, I would like to acknowledge all my students, past and present, whose comments and questions have helped me focus on the nature, meaning, and essentials of Earth science. A special debt of gratitude goes to Dan Kaveney and his Prentice Hall colleagues who skillfully guided this manuscript from its inception to completion. They are true professionals with whom I feel fortunate to be associated.

Thanks,
Ken Pinzke

I hear and I forget.

I see and I remember.

I do and I understand.
Ancient Proverb

Introduction

The *Introduction* opens with a brief discussion of Earth's major "spheres." A definition of Earth science is followed by a discussion of the four areas that are traditionally included—geology, oceanography, meteorology, and astronomy. The importance of understanding basic Earth science principles when examining resources and environmental issues is also presented. The Introduction closes with a discussion of the nature of scientific inquiry.

Learning Objectives

After reading, studying, and discussing the Introduction, you should be able to:

- Describe Earth's four "spheres."
- List the sciences traditionally included in Earth science.
- Discuss some resource and environmental issues.
- Describe the nature of scientific inquiry.

Introduction Review

- The four "spheres" of Earth include the 1) *hydrosphere*, the dynamic mass of water that is continually on the move; 2) *atmosphere*, the gaseous envelope surrounding Earth; 3) *solid Earth*, which is divided into the *core*, *mantle*, and *crust*; and 4) *biosphere*, which includes all life.

- *Earth science* is the name for all the sciences that collectively seek to understand Earth and its neighbors in space. It includes geology, oceanography, meteorology, and astronomy.

- *Environment* refers to everything that surrounds and influences an organism. These influences can be biological, social, or physical. *Resources* are an important environmental concern. The two broad categories of resources are 1) *renewable*, which means that they can be replenished over relatively short time spans; and 2) *nonrenewable*.

- Environmental problems can be local, regional, or global. Human-induced problems include urban air pollution, acid rain, ozone depletion, and global warming. Natural hazards imposed by the physical environment include earthquakes, landslides, floods, and hurricanes. As world population continues to grow, pressures on the environment increase as well.

- All science is based on the assumption that the natural world behaves in a consistent and predictable manner. The process through which scientists gather facts through observations and formulate scientific *hypotheses* and *theories* is called the *scientific method*. To help determine what is occurring in the natural world, scientists often 1) collect facts, 2) develop a scientific hypothesis, 3) construct experiments to validate the hypothesis, and 4) accept, modify, or reject the hypothesis on the basis of extensive testing. Other discoveries represent purely theoretical ideas which have stood up to extensive examination. Still other scientific advancements have been made when a totally unexpected happening occurred during an experiment.

Chapter Outline

I. Four "spheres" of the Earth
  A. Hydrosphere
    1. Ocean the most prominent feature
      a. Nearly 71% of Earth's surface
      b. About 97% of Earth's water
    2. Also includes fresh water
  B. Atmosphere
  C. Solid Earth
    1. Core
    2. Mantle
    3. Crust
  D. Biosphere
    1. Includes all life
    2. Influences other three spheres
II. Earth science
  A. Encompasses all sciences that seek to understand
    1. Earth
    2. Earth's neighbors in space
  B. Includes
    1. Geology
      a. Physical geology
      b. Historical geology
    2. Oceanography
      a. Not a separate and distinct science
      b. Integrates
        1. Chemistry
        2. Physics
        3. Geology
        4. Biology
    3. Meteorology
    4. Astronomy
III. Resources and environmental issues
  A. Environment
    1. Surrounds and influences organisms
    2. Influences on organisms
      a. Biological
      b. Social
      c. Physical
  B. Resources
    1. Important environmental concern
    2. Include
      a. Water
      b. Soil
      c. Minerals
      d. Energy
    3. Two broad categories
      a. Renewable
        1. Can be replenished
        2. Examples
          a. Plants
          b. Wind energy
      b. Nonrenewable
        1. Fixed quantities
        2. Examples
          a. Metals
          b. Fuels
  C. Environmental problems
    1. Local, regional, and global
    2. Human-induced and accentuated
      a. Urban air pollution
      b. Acid rain
      c. Ozone depletion
      d. Global warming
    3. Natural hazards
      a. Earthquakes
      b. Landslides
      c. Floods
      d. Hurricanes
    4. World population pressures
IV. Scientific inquiry
  A. Science assumes the natural world is
    1. Consistent
    2. Predictable
  B. Goal of science
    1. To discover patterns in nature, and
    2. To use the knowledge to predict
  C. An idea can become a
    1. Hypothesis (untested)
    2. Theory (tested and confirmed)
  D. Scientific method
    1. Gather facts through observation
    2. Formulate
      a. Hypotheses
      b. Theories
  E. Scientific knowledge is gained through
    1. Following systematic steps
      a. Collecting facts
      b. Developing a hypothesis (untested)
      c. Conduct experiments
      d. Reexamine the hypothesis and
        1. Accept
        2. Modify
        3. Reject
    2. Theories that withstand examination
    3. Totally unexpected occurrences

Vocabulary Review

Choosing from the list of key terms, furnish the most appropriate response for the following statements.

1. The solid Earth is divided into three principal units: the very dense ____________; the less-dense ____________; and the ____________.

2. In scientific inquiry, a preliminary untested explanation is a scientific ____________.

3. The science of ____________ is traditionally divided into two parts—physical and historical.

4. ____________ is the study of the atmosphere and the processes that produce weather and climate.

5. Water, soil, minerals, and energy are each considered an important ____________.

6. A scientific ____________ is a well-tested and widely accepted view that scientists agree best explains certain observable

7. The dynamic water portion of Earth's physical environment is the ____________.

8. The science that involves the application of all sciences in a comprehensive and interrelated study of the oceans in all their aspects and relationships is called ____________.

9. A(n) ____________ is a resource that can be replenished over a relatively short time span.

10. The "sphere" of Earth, called the ____________, includes all life.

11. A resource, such as oil, that cannot be replenished is classified as a(n) ____________.

12. ____________ is the name for all the sciences that collectively seek to understand Earth and its neighbors in space.

13. Everything that surrounds and influences an organism is termed the ____________.

14. ____________ is the study of the universe.

15. The life-giving gaseous envelope surrounding Earth is the ____________.

Key Terms

Page numbers shown in ( ) refer to the textbook page where the term first appears.

astronomy (p. 4)
atmosphere (p. 1)
biosphere (p. 2)
core (p. 2)
crust (p. 2)
Earth science (p. 2)
environment (p. 4)
geology (p. 2)
hydrosphere (p. 1)
hypothesis (p. 6)
mantle (p. 2)
meteorology (p. 3)
nonrenewable resource (p. 5)
oceanography (p. 3)
renewable resource (p. 4)
resource (p. 4)
theory (p. 7)

Comprehensive Review

1. What are the sciences that are traditionally included in Earth science?

2. Describe the difference between renewable and nonrenewable resources.

3. List and briefly describe Earth's four "spheres."
   1)
   2)
   3)
   4)

4. How would you explain to a friend what the science of Earth science involves?

5. What is the difference between a scientific theory and a scientific hypothesis?

6. What are the four steps that are often used in science to gain knowledge?
   1)
   2)
   3)
   4)

7. Label each of the four units of the solid Earth at the appropriate letter in Figure I.1.

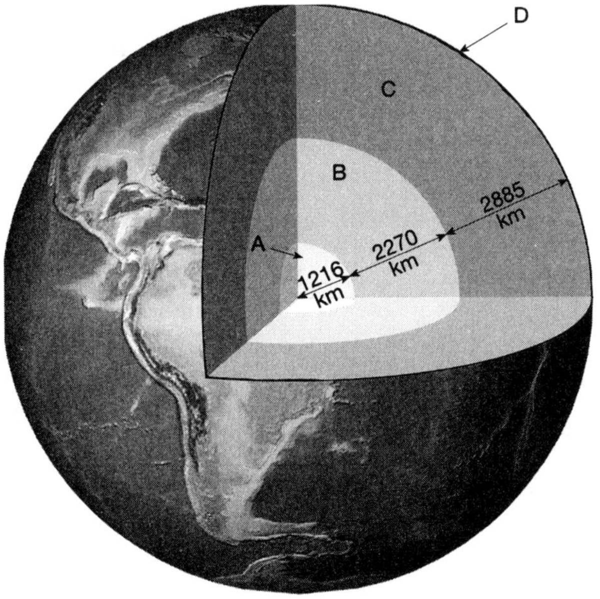

Figure I.1

8. Describe the following two broad areas of geology:
   a) Physical geology:

   b) Historical geology:

Practice Test

Multiple choice. Choose the best answer for the following multiple choice questions.

1. The ____________ strongly influences the other three "spheres" because without life their makeups and natures would be much different.

   a) atmosphere   b) hydrosphere   c) solid Earth   d) biosphere

2. Earth science attempts to relate our planet to the larger universe by investigating the science of ____________.

   a) geology   c) meteorology   e) biology
   b) oceanography   d) astronomy

3. The science that studies the processes that produce weather and climate is ____________.

   a) geology   c) meteorology   e) biology
   b) oceanography   d) astronomy

4. The global ocean covers ____________ percent of Earth's surface.

   a) 71   c) 42   e) 83
   b) 38   d) 65

5. A scientific ____________ is a preliminary untested explanation which tries to explain how or why things happen in the manner observed.

   a) estimate   c) fact   e) idea
   b) law   d) hypothesis

6. The Earth "sphere," called the ____________, includes the fresh water found in streams, lakes, glaciers, as well as that existing underground.

   a) atmosphere   b) hydrosphere   c) biosphere   d) solid Earth

7. The science that includes the study of the composition and movements of seawater, as well as coastal processes, sea floor topography, and marine life is ____________.

   a) geology   c) meteorology   e) biology
   b) oceanography   d) astronomy

8. Which one of the following is NOT an example of a nonrenewable resource?

   a) solar energy   c) copper   e) aluminum
   b) natural gas   d) coal

9. An understanding of Earth is essential for ____________.

   a) location and recovery of basic resources
   b) minimizing the effects of natural hazards
   c) dealing with the human impact on the environment
   d) all the above

10. This science is divided into two broad divisions—physical and historical.

   a) geology   c) meteorology   e) biology
   b) oceanography   d) astronomy

11. The annual per capita consumption of metallic and nonmetallic mineral resources for the United States is nearly ____________ tons.
    a) 2 c) 8 e) 15
    b) 6 d) 11

12. The conditions that surround and influence an organism are referred to as the ____________.
    a) resource b) social sphere c) technosphere d) environment

13. Which one of the following is NOT an example of a renewable resource?
    a) water energy c) iron e) chickens
    b) lumber d) cotton

14. In scientific inquiry, when competing hypotheses have been eliminated, a hypothesis may be elevated to the status of a scientific ____________.
    a) estimate c) idea e) truth
    b) theory d) understanding

15. The science whose name literally means "study of Earth" is ____________.
    a) geology c) meteorology e) biology
    b) oceanography d) astronomy

16. Scientific advancements can be made ____________.
    a) by applying systematic steps
    b) by accepting purely theoretical ideas which stand up to extensive investigation
    c) from unexpected happenings during an experiment
    d) all the above

17. Earth is approximately ____________ billion years old.
    a) 2.5 c) 4.6 e) 9.8
    b) 3.0 d) 6.2

True/false. For the following true/false questions, if a statement is not completely true, mark it false. For each false statement, change the ***italicized*** *word to correct the statement.*

1. ___ The life-giving gaseous envelope surrounding Earth is called the *atmosphere.*

2. ___ The *biosphere* is a dynamic mass of water that is continually on the move from the oceans to the atmosphere, precipitating back to the land, and running back to the ocean.

3. ___ Although some *renewable* resources such as aluminum can be used over and over again, others, such as oil, cannot be recycled.

4. ___ The aim of *physical geology* is to understand the origin of Earth and the development of the planet since its formation.

5. ___ Science is based on the assumption that the natural world behaves in a(n) *unpredictable* manner.

6. ___ *Biosphere* refers to everything that surrounds and influences an organism.

7. ___ To determine what is happening in the natural world, scientists collect *facts* through observation and measurement.

8. ___ The light and very thin outer skin of Earth is called the *mantle.*

9. ___ All life on Earth is included in the *biosphere.*

10. ___ A scientific *theory* is a well-tested and widely accepted view that scientists agree best explains certain observable facts.

Written questions

1. List the four steps that scientists often use to conduct experiments and gain scientific knowledge.

2. List and briefly describe Earth's four "spheres."

3. Define Earth science. List the four areas commonly included in Earth science.

MINERALS:
Building Blocks of Rocks

1

Minerals: Building Blocks of Rocks begins with an explanation of the difference between minerals and rocks, followed by a formal definition of a mineral. Elements, atoms, compounds, ions, and atomic bonding are discussed. Also investigated are isotopes and radioactivity. Following descriptions of the properties used in mineral identification, the silicate and nonsilicate mineral groups are examined. The chapter concludes with a presentation of mineral resources, reserves, and ores.

Learning Objectives

After reading, studying, and discussing this chapter, you should be able to:

- Explain the difference between a mineral and a rock.
- Describe the basic structure of an atoms and explain how atoms combine.
- List the most important elements that compose Earth's continental crust.
- Explain isotopes and radioactivity.
- Describe the physical properties of minerals and how they can be used for mineral identification.
- List the basic compositions and structures of the silicate minerals.
- List the economic use of some nonsilicate minerals.
- Distinguish between mineral resources, reserves, and ores.

Chapter Review

- A *mineral* is a naturally occurring inorganic solid that possesses a definite chemical structure, which gives it a unique set of physical properties. Most *rocks* are aggregates composed of two or more minerals.
- The building blocks of minerals are *elements.* An *atom* is the smallest particle of matter that still retains the characteristics of an element. Each atom has a *nucleus* which contains *protons* and *neutrons.* Orbiting the nucleus of an atom are *electrons.* The number of protons in an atom's nucleus determines its *atomic number* and the name of the element. Atoms bond together to form a *compound* by either gaining, losing, or sharing electrons with another atom.
- *Isotopes* are variants of the same element, but with a different *mass number* (the total number of neutrons plus protons found in an atom's nucleus). Some isotopes are unstable and disintegrate naturally through a process called *radioactive decay.*

• The properties of minerals include *crystal form, luster, color, streak, hardness, cleavage, fracture,* and *specific gravity*. In addition, a number of special physical and chemical properties (*taste, smell, elasticity, malleability, feel, magnetism, double refraction,* and *chemical reaction to hydrochloric acid*) are useful in identifying certain minerals. Each mineral has a unique set of properties which can be used for identification.

• The eight most abundant elements found in Earth's continental crust (oxygen, silicon, aluminum, iron, calcium, sodium, potassium, and magnesium) also compose the majority of minerals.

• The most common mineral group is the *silicates*. All silicate minerals have the *silicon-oxygen tetrahedron* as their fundamental building block. In some silicate minerals the tetrahedra are joined in chains; in others, the tetrahedra are arranged into sheets, or three-dimensional networks. Each silicate mineral has a structure and a chemical composition that indicates the conditions under which it was formed.

• The *nonsilicate* mineral groups include the *oxides* (e.g., magnetite, mined for iron), *sulfides* (e.g., sphalerite, mined for zinc), *sulfates* (e.g., gypsum, used in plaster and frequently found in sedimentary rocks), *native elements* (e.g., graphite, a dry lubricant), *halides* (e.g., halite, common salt and frequently found in sedimentary rocks), and *carbonates* (e.g., calcite, used in portland cement and a major constituent in two well-known rocks: limestone and marble).

• The term *ore* is used to denote useful metallic minerals, like hematite (mined for iron) and galena (mined for lead), that can be mined for a profit, as well as some nonmetallic minerals, such as fluorite and sulfur, that contain useful substances.

Chapter Outline

I. Minerals versus rocks
   A. Minerals
      1. Naturally occurring
      2. Inorganic
      3. Solid
      4. Definite chemical structure
   B. Nearly 4000 known minerals
   C. Rocks—aggregates (mixtures) of minerals
II. Composition and structure of minerals
   A. Elements
      1. Basic building blocks of minerals
      2. Over 112 are known
   B. Atoms
      1. Smallest particles of matter
      2. Have all characteristics of an element
III. How atoms are constructed
   A. Nucleus
      1. Contains protons—positive electrical charge
      2. Contains neutrons—neutral electrical charge
   B. Energy levels, or shells
      1. Surround nucleus
      2. Contain electrons—negative electrical charges
   C. Atomic number—number of protons in atom's nucleus
   D. Bonding of atoms
      1. Forms compound—two or more elements
      2. Ions—atoms that gain or lose electrons
   E. Isotopes
      1. Have varying number of neutrons
      2. Have different mass numbers—sum of neutrons plus protons
      3. Many are radioactive—emit energy and particles
IV. Minerals
   A. Properties of minerals
      1. Crystal form
      2. Luster

3. Color
4. Streak
5. Hardness
6. Cleavage
7. Fracture
8. Specific gravity
9. Other
   a. Taste
   b. Smell
   c. Elasticity
   d. Malleability
   e. Feel
   f. Magnetism
   g. Double refraction
   h. Reaction to hydrochloric acid

B. A few dozen are called rock-forming-minerals
   1. Eight elements compose most rock-forming minerals [oxygen (O), silicon (Si), aluminum(Al), iron (Fe), calcium (Ca), potassium (K), magnesium (Mg)]
   2. Most abundant atoms in Earth's crust are
      a. Oxygen (46.6% of crust by weight)
      b. Silicon (27.7% of crust by weight)

C. Mineral groups
   1. Silicate minerals
      a. Most common mineral group
      b. Contain silicon-oxygen tetrahedron
      c. Groups based upon tetrahedron arrangement
         1. Olivine—independent tetrahedron
         2. Pyroxene group—arranged in chains
         3. Amphibole group—arranged in double chains
         4. Micas
            a. Arranged in sheets
            b. Types
               1. Biotite
               2. Muscovite
         5. Feldspars
            a. Three-dimensional network
            b. Types
               1. Orthoclase
               2. Plagioclase
         6. Quartz—three-dimensional network
      d. Feldspars most plentiful group
      e. Crystallize from molten material
   2. Nonsilicate minerals
      a. Major groups
         1. Oxides
         2. Sulfides
         3. Sulfates
         4. Halides
         5. Carbonates
         6. "Native" elements
      b. Carbonates
         1. Major rock-forming group
         2. Found in limestone and marble
      c. Halite and gypsum—Found in sedimentary rocks
      d. Many have economic value

D. Mineral resources
   1. Reserves
   2. Ores—Can be mined at a profit
   3. Economic factors may change

Vocabulary Review

Choosing from the list of key terms, furnish the most appropriate response for the following statements.

1. A(n) ____________ is an electrically neutral subatomic particle found in the nucleus of an atom.
2. The smallest part of an element that still retains the element's properties is a(n) ____________.
3. A(n) ____________ is a useful metallic mineral that can be mined at a profit.

4. A naturally occurring, inorganic solid that possesses a definite chemical structure, which gives it a unique set of physical properties is a(n) ____________.

5. A(n) ____________ is an atom that has an electric charge because of a gain or loss of electrons.

6. The tendency of a mineral to break along planes of weak bonding is the property called ____________.

7. The subatomic particle that contributes mass and a positive electrical charge to an atom is a(n) ____________.

8. A(n) ____________ is composed of two or more elements bonded together in definite proportions.

Key Terms

Page numbers shown in ( ) refer to the textbook page where the term first appears.

atom (p. 14)
atomic number (p. 14)
cleavage (p. 17)
color (p. 16)
compound (p. 14)
crystal form (p. 16)
electron (p. 14)
element (p. 13)
energy level (shells) (p. 14)
fracture (p. 18)
hardness (p. 17)
ion (p. 15)
isotope (p. 15)
luster (p. 16)
mass number (p. 15)
mineral (p. 13)
mineral resource (p. 22)
Mohs scale (p. 17)
neutron (p. 14)
nucleus (p. 14)
ore (p. 22)
proton (p. 14)
radioactive decay (p. 16)
reserve (p. 22)
rock (p. 12)
silicates (p. 19)
silicon-oxygen tetrahedron (p. 20)
specific gravity (p. 18)
streak (p. 17)

9. Silicon and oxygen combine to form the framework of the most common mineral group, the ____________.

10. An already identified deposit from which minerals can be extracted profitably is called a(n) ____________.

11. A(n) ____________ is a subatomic particle with a negative electrical charge.

12. A(n) ____________ can be defined simply as an aggregate of minerals.

13. Each atom has a central region called the ____________.

14. The number of protons in an atom's nucleus determines its ____________ and the name of the element.

15. A(n) ____________ is a large collection of electrically neutral atoms, all having the same atomic number.

16. The sum of the neutrons and protons in the nucleus is the atom's ____________.

17. Electrons are located at a given distance from an atom's nucleus in a region called the ____________.

18. A variation of the same element but with a different mass number is called a(n) ____________.

19. The external expression of the internal orderly arrangement of atoms is called the mineral's ____________.

20. The process called ____________ involves the natural disintegration of unstable isotopes.

21. The property of a mineral which involves the appearance or quality of light reflecte from its surface is termed ____________.
22. The fundamental building block of all silicate minerals is the ____________.
23. A(n) ____________ is a mineral that can be recovered for a useful purpose.
24. The property called ____________ is the color of a mineral in the powdered form.
25. The resistance of a mineral to abrasion or scratching, called ____________, is one of the most useful diagnostic properties.
26. Minerals that do not exhibit cleavage when broken are said to ____________.
27. Comparing the weight of a mineral to the weight of an equal volume of water determines the mineral's ____________.
28. To assign a value to a mineral's resistance to abrasion or scratching, geologists use a standard scale called ____________.
29. Due to impurities, a mineral's ____________ is often an unreliable diagnostic property.

Comprehensive Review

1. List the four characteristics that any Earth material must exhibit in order to be considered a mineral.

    1)

    2)

    3)

    4)

2. List the three main subatomic particles found in an atom and describe how they differ from one another. Also indicate where each is located in an atom by selecting the proper letter in Figure 1.1.

    1) Letter: ____

    2) Letter: ____

    3) Letter: ____

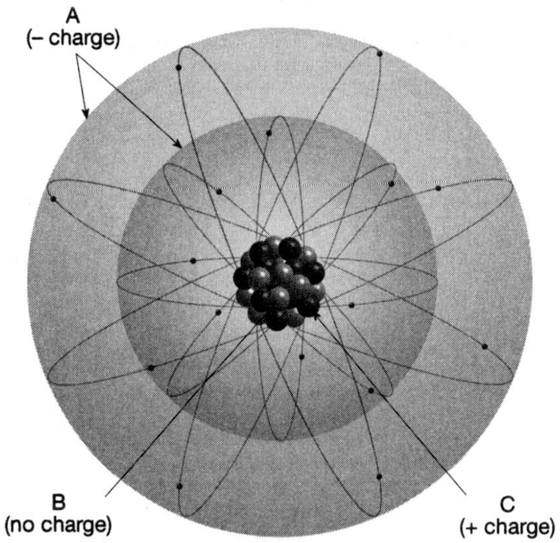

Figure 1.1

3. Briefly explain the difference between a mineral and a rock.
4. If an atom has 6 protons, 6 electrons, and 8 neutrons, what is the atom's

   a) Atomic number? ____________

   b) Mass number? ____________

5. If an atom has 17 electrons and its mass number is 35, calculate the following:

   a) Number of protons ____________

   b) Atomic number ____________

   c) Number of neutrons ____________

6. Briefly explain the formation of the chemical bond involving a sodium atom and chlorine atom to produce the compound, sodium chloride.

7. Examine the diagram of an atom shown in Figure 1.2A. The atom's nucleus contains 8 protons and 5 neutrons. Is the atom shown in diagram 1.2A an ion? Explain your answer.

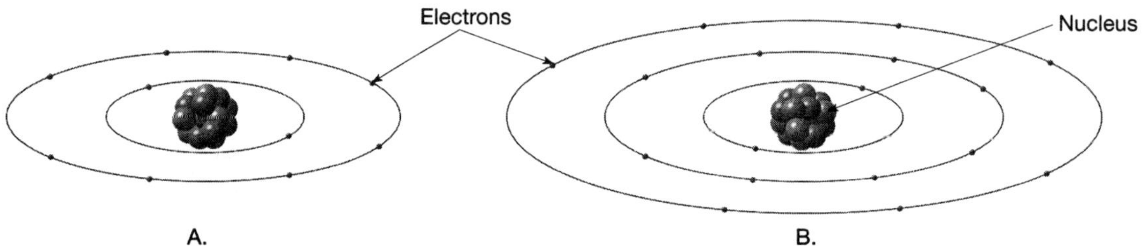

Figure 1.2

   a) What is the mass number of the atom shown in Figure 1.2A? ____________

   b) What is the atomic number of the atom shown in Figure 1.2A? ____________

8. Examine the diagram of an atom shown in Figure 1.2B. The atom's nucleus contains 17 protons. Is the atom shown in diagram 1.2B an ion? Explain your answer.

9. Define a mineral.

10. In what way does an isotope vary from the common form of the same element?

11. Briefly describe each of the following properties of minerals.

    a) Luster:

    b) Crystal form:

    c) Streak:

    d) Hardness:

    e) Cleavage:

    f) Fracture:

    g) Specific gravity:

12. What are the two most common elements that compose Earth's continental crust? Also list the approximate percentage (by weight) of the continental crust that each element composes, along with its chemical symbol.

13. Figure 1.3 represents the silicon-oxygen tetrahedron.

    Letter A identifies ____________ atoms,

    while letter B is a ____________ atom.

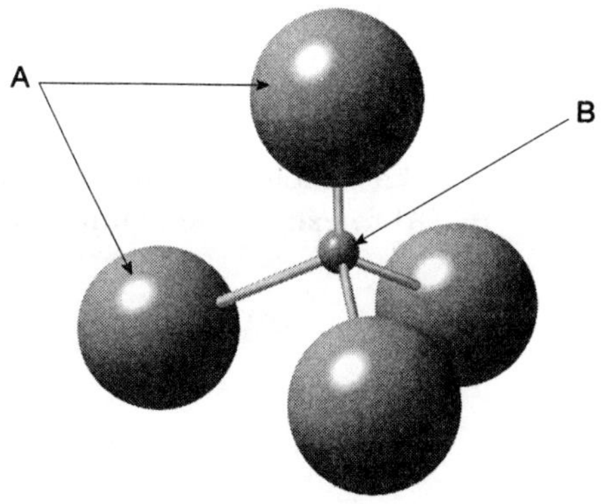

Figure 1.3

14. Name the silicate mineral group(s) that exhibits each of the following silicate structures.

    a) Three-dimensional networks:

    b) Sheets:

    c) Double chains:

    d) Single chains:

    e) Single tetrahedron:

15. Name three common rock-forming silicate minerals and the most common rock-forming nonsilicate mineral.

    a) Three common rock-forming silicate minerals:

    b) Most common rock-forming nonsilicate mineral:

16. Define a rock.

17. How do most silicate minerals form?

18. List a mineral that is an ore of each of the following elements.

    a) Mercury:

    b) Lead:

    c) Zinc:

    d) Iron:

19. Briefly explain what happens to an isotope as it decays.

20. Could a sample of Earth material that contains 10 percent aluminum by weight be profitably mined to extract the aluminum? Explain your answer.

Practice Test

Multiple choice. Choose the best answer for the following multiple choice questions.

1. Which one of the following is NOT a silicate mineral group?
   a) micas
   b) feldspars
   c) carbonates
   d) pyroxenes
   e) quartz

2. The basic building block of the silicate minerals ____________.
   a) has the shape of a cube
   b) contains 1 silicon atom and 4 oxygen atoms
   c) always occurs independently
   d) contains 2 iron atoms for each silicon atom
   e) contains 1 oxygen atom and 4 silicon atoms

3. Which one of the following is NOT one of the eight most abundant elements in Earth's continental crust?
   a) oxygen
   b) iron
   c) aluminum
   d) calcium
   e) hydrogen

4. Atoms of the same element possess the same number of ____________.
   a) isotopes
   b) ions
   c) protons
   d) neutrons
   e) compounds

5. Atoms containing the same numbers of protons and different numbers of neutrons are ____________.
   a) isotopes
   b) ions
   c) protons
   d) neutrons
   e) compounds

6. Which subatomic particle is most involved in chemical bonding?
   a) electrons
   b) ions
   c) protons
   d) neutrons
   e) nucleus

7. The number of ____________ in an atom's nucleus determines which element it is.
   a) electrons
   b) ions
   c) protons
   d) neutrons
   e) isotopes

8. Oxygen comprises about __________ percent by weight of Earth's continental crust.
   a) 12
   b) 27
   c) 47
   d) 63
   e) 85

9. The smallest particle of an element that still retains all the element's properties is a(n) ____________.
   a) compound
   b) rock
   c) atom
   d) neutron
   e) isotope

10. Atoms that gain or lose electrons become ____________.
    a) electrons
    b) ions
    c) protons
    d) neutrons
    e) compounds

11. When two or more elements bond together in definite proportions they form a(n) ____________.
    a) rock
    b) ion
    c) atom
    d) compound
    e) nucleus

12. The sum of the neutrons and protons in an atom's nucleus is the atom's ___________.
    a) specific gravity c) atomic number e) isotope number
    b) atomic mass d) energy-levels

13. Radioactive decay occurs when ___________.
    a) unstable nuclei disintegrate
    b) ions form
    c) a mineral is cleaved
    d) electrons change energy-levels
    e) atoms bond

14. The property which is a measure of the resistance of a mineral to abrasion or scratching is ___________.
    a) crystal form c) cleavage e) hardness
    b) streak d) fracture

15. The property which involves the color of a mineral in its powdered form is ___________.
    a) crystal form c) cleavage e) hardness
    b) streak d) fracture

16. The tendency of a mineral to break along planes of weak bonding is called ___________.
    a) luster c) cleavage e) bonding
    b) crystal form d) fracture

17. Minerals that break into smooth curved surfaces, like those seen in broken glass, have a unique type of fracture, called _________ fracture.
    a) conchoidal c) rounded e) uneven
    b) irregular d) vitreous

18. The basic building block of the silicate minerals is the silicon-oxygen ___________.
    a) cube c) octagon e) sphere
    b) tetrahedron d) rhombohedron

19. Which of the following is NOT a primary element that joins silicate structures?
    a) calcium c) potassium e) oxygen
    b) iron d) magnesium

20. Most silicate minerals form from ___________.
    a) other minerals c) erosion e) water
    b) radioactive decay d) molten rock

21. Which silicate mineral group has a double chain silicate structure?
    a) pyroxenes c) micas e) quartz
    b) amphiboles d) feldspars

22. Which major rock-forming mineral is the chief constituent of the rocks limestone and marble?
    a) gypsum c) mica e) calcite
    b) halite d) feldspar

23. The scale used by geologists to measure the hardness of a mineral is called _________ scale.
    a) Mohs c) Richters e) Playfairs
    b) Smiths d) Bowens

24. Which nonsilicate mineral group contains the ores of iron?
    a) oxides c) sulfates e) carbonates
    b) sulfides d) halides

25. Which one of the following statements concerning minerals is NOT true?
    a) naturally occurring c) organic e) formed of atoms
    b) solid d) definite chemical structure

True/false. For the following true/false questions, if a statement is not completely true, mark it false. For each false statement, change the ***italicized*** *word to correct the statement.*

1. ___ Ions form when atoms gain or lose *neutrons.*
2. ___ *Protons* have an opposite electrical charge from electrons.
3. ___ Nearly *eighty* minerals have been named.
4. ___ Only *eight* elements compose the bulk of the rock-forming minerals.
5. ___ The elements silicon and *carbon* comprise nearly three-fourths of Earth's continental crust.
6. ___ The most common mineral group is the *silicate* group.
7. ___ The structure of the silicon-oxygen tetrahedron consists of *four* oxygen atoms surrounding a smaller silicon atom.
8. ___ Each silicate mineral group has a particular silicate *structure.*
9. ___ Silicate minerals tend to cleave *through* the silicon-oxygen structures.
10. ___ Each *silicate* mineral has a structure and chemical composition that indicate the conditions under which it formed.
11. ___ Nonsilicate minerals make up about *one-tenth* of the continental crust.
12. ___ The mineral calcite is a *sulfide* mineral.
13. ___ *Quartz* is the mineral from which plaster and other similar building materials are composed.
14. ___ *Reserves* are already-identified mineral deposits from which minerals can be extracted profitably.
15. ___ *Luster* is the external expression of a mineral's internal orderly arrangement of atoms.
16. ___ Cleavage is defined by the number of planes exhibited and the *angles* at which they meet.
17. ___ Vitreous, pearly, and silky are types of *metallic* lusters.
18. ___ Isotopes of the same element have *different* mass numbers.
19. ___ An *ion* is composed of two or more elements bonded together.
20. ___ The number of *protons* in an atom's nucleus determines its atomic number.
21. ___ Uncombined atoms have the same number of *neutrons* as protons.
22. ___ Individual *electrons* are located at given distances from an atom's nucleus in regions called energy levels.

23. ___ A(n) *compound* is the smallest particle of matter that has all the characteristics of an element.

24. ___ A mineral is a naturally occurring *inorganic* solid that possesses a definite chemical structure.

25. ___ Quartz is *softer* than calcite.

Written questions

1. What are the three main particles of an atom? How do they differ from one another?

2. Why don't most mineral samples demonstrate visibly their crystal form?

3. Describe the basic building block of all silicate minerals.

4. Define a mineral.

5. Explain the difference between a mineral resource and a mineral reserve.

ROCKS:

Materials of the Lithosphere

2

Rocks: Materials of the Lithosphere opens with a discussion of the rock cycle, which presents a general overview of the origins and processes involved in forming the three major rock groups—igneous rock, sedimentary rock, and metamorphic rock. A discussion of the crystallization of magma is followed by an examination of the classification, textures, and compositions of igneous rocks. After presenting the processes of mechanical and chemical weathering, the classification of sedimentary rocks, as well as some of their common features, is discussed. The chapter concludes with an investigation of the agents of metamorphism, the textural and mineralogical changes that take place during metamorphism, and some common metamorphic rocks.

Learning Objectives

After reading, studying, and discussing this chapter, you should be able to:

- Diagram and discuss the rock cycle.
- List the geologic processes involved in the formation of each rock group.
- Briefly explain crystallization of magma.
- List the criteria used to classify igneous rocks.
- List the names, textures, and environments of formation for the most common igneous rocks.
- Explain the processes of mechanical and chemical weathering.
- List the major sources of materials that accumulate as sediment.
- List the criteria used to classify sedimentary rocks.
- Explain the difference between detrital and chemical sedimentary rocks.
- List the names, textures, and environments of formation for the most common sedimentary rocks.
- List the common features of sedimentary rocks.
- Describe the agents of metamorphism.
- List the criteria used to classify metamorphic rocks.
- List the names, textures, and environments of formation for the most common metamorphic rocks.

Chapter Review

- *Igneous rock* forms from magma that cools and solidifies in a process called *crystallization*. *Sedimentary rock* forms from the *lithification* of *sediment*. *Metamorphic rock* forms from rock that has been subjected to great pressure and heat in a process called *metamorphism*.

- The rate of cooling of magma greatly influences the size of mineral crystals in igneous rock. The four basic igneous rock textures are 1) *fine-grained*, 2) *coarse-grained*, 3) *porphyritic*, and 4) *glassy*.

• The mineral makeup of an igneous rock is ultimately determined by the chemical composition of the magma from which it crystallized. N.L. Bowen showed that as magma cools, minerals crystallize in an orderly fashion. Crystal settling can change the composition of magma and cause more than one rock type to form from a common parent magma.

• Igneous rocks are classified by their texture and mineral composition.

• Weathering is the response of surface materials to a changing environment. Mechanical weathering, the physical disintegration of material into smaller fragments, is accomplished by frost wedging, expansion resulting from unloading, and biological activity. Chemical weathering involves processes by which the internal structures of minerals are altered by the removal and/or addition of elements. It occurs when materials are oxidized or react with acid, such as carbonic acid.

• Detrital sediments are materials that originate and are transported as solid particles derived from weathering. Chemical sediments are soluble materials produced largely by chemical weathering that are precipitated by either inorganic or organic processes. Detrital sedimentary rocks, which are classified by particle size, contain a variety of mineral and rock fragments, with clay minerals and quartz the chief constituents. Chemical sedimentary rocks often contain the products of biological processes such as shells or mineral crystals that form as water evaporates and minerals precipitate. Lithification refers to the processes by which sediments are transformed into solid sedimentary rocks.

• Common detrital sedimentary rocks include shale (the most common sedimentary rock), sandstone, and conglomerate. The most abundant chemical sedimentary rock is limestone, composed chiefly of the mineral calcite. Rock gypsum and rock salt are chemical rocks that form as water evaporates and triggers the deposition of chemical precipitates.

• Some of the features of sedimentary rocks that are often used in the interpretation of Earth history and past environments include strata, or beds (the single most characteristic feature), bedding planes, and fossils.

• Two types of metamorphism are 1) regional metamorphism and 2) contact metamorphism. The agents of metamorphism include heat, pressure, and chemically active fluids. Heat is perhaps the most important because it provides the energy to drive the reactions that result in the recrystallization of minerals. Metamorphic processes cause many changes in rocks, including increased density, growth of larger mineral crystals, reorientation of the mineral grains into a layered or banded appearance known as foliation, and the formation of new minerals.

• Some common metamorphic rocks with a foliated texture include slate, schist, and gneiss. Metamorphic rocks with a nonfoliated texture include marble and quartzite.

Chapter Outline

I. Rock cycle
  A. Shows relations among the three rock types
  B. Proposed by James Hutton in late 1700s
  C. The cycle
    1. Magma
    2. Crystallization
    3. Igneous rock
    4. Weathering
    5. Transportation
    6. Deposition
    7. Sediment
    8. Lithification
    9. Sedimentary rock
    10. Metamorphism
    11. Metamorphic rock
    12. Melting
    13. Magma

- D. Full cycle does not always take place
  1. Sedimentary rock melts
  2. Igneous rock metamorphosed
  3. Sedimentary rock weathers
  4. Metamorphic rock weathers

II. Igneous rocks
- A. Form as magma cools and crystallizes
  1. Rocks formed inside Earth are called plutonic or intrusive rocks
  2. Rocks formed on surface
     - a. Formed from lava which is similar to magma, but no gas
     - b. Called volcanic or extrusive rocks
- B. Crystallization of magma
  1. Ions arranged into orderly patterns
  2. Crystal size is determined by rate of cooling
     - a. Slow rate forms large crystals
     - b. Fast rate forms microscopic crystals
     - c. Very fast rate forms glass
- C. Classification
  1. Uses texture and mineral composition
     - a. Texture
       1. Size and arrangement of crystals
       2. Types
          - a. Fine-grained
          - b. Coarse-grained
          - c. Porphyritic (two crystal sizes)
          - d. Glassy
     - b. Mineral composition
       1. Explained by Bowen's reaction series that shows the order of mineral crystallization
       2. Influenced by crystal settling in magma
- D. Naming igneous rocks
  1. Basaltic rocks
     - a. Derived from first minerals to crystallize
     - b. Rich in iron and magnesium
     - c. Low in silica
     - d. Common rock is basalt
  2. Granitic rocks
     - a. From last minerals to crystallize
     - b. Mainly feldspar and quartz
     - c. High silica
     - d. Common rock is granite

III. Weathering
- A. Response to a changing environment
- B. Two types
  1. Mechanical weathering
     - a. Material broken into smaller pieces
     - b. Processes
       1. Frost wedging
       2. Expansion from unloading
       3. Biological activity
  2. Chemical weathering
     - a. Alters internal structure of minerals
     - b. Water most important agent
       1. Oxygen will oxidize materials
       2. Acidic water
          - a. Carbon dioxide in water makes carbonic acid
          - b. Destroys a mineral's structure
     - c. Weathering of granite
       1. Materials produced from feldspar
          - a. Clay (most abundant)
          - b. Potassium ions (dissolved)
          - c. Silica (dissolved)
       2. Quartz
          - a. Does not chemically weather
          - b. Large grains become small grains

IV. Sedimentary rocks
- A. Form from sediment
  1. Weathered products
- B. Form about 75% of outcrops on continents
- C. Used to construct much of Earth's history
  1. Clues to past environments
  2. Information about sediment transport
  3. Fossils
- D. Economic importance
  1. Coal
  2. Petroleum and natural gas
  3. Sources of iron and aluminum
- E. Classification
  1. Two groups based on source of material
     - a. Detrital rocks
       1. Material is solid particles
       2. Classified by particle size
       3. Common rocks
          - a. Shale (most abundant)
          - b. Sandstone
          - c. Conglomerate
          - d. Siltstone
     - b. Chemical rocks
       1. Derived from material once in solution

2. Can precipitate from water to form sediment
   a. Directly
   b. Through life processes (biochemical origin)
2. Common rocks
   a. Limestone (most abundant)
   b. Travertine
   c. Microcrystalline quartz
      1. Chert
      2. Flint
      3. Jasper
      4. Agate
   d. Evaporites
      1. Rock salt
      2. Gypsum
   e. Coal
      1. Lignite
      2. Bituminous

F. Produced through lithification
1. Loose sediments transformed into solid rock
2. Lithification processes
   a. Compaction
   b. Cementation by the materials
      1. Calcite
      2. Silica
      3. Iron oxide

G. Features
1. Strata, or beds (most characteristic)
2. Bedding planes that separate strata
3. Fossils
   a. Traces or remains of prehistoric life
   b. Comprise most important inclusions
   c. Help determine past environments
   d. Used as time indicators
   e. Used for matching rocks from different places

V. Metamorphic rocks

A. "Changed form" rocks

B. Can form from
1. Igneous rocks
2. Sedimentary rocks
3. Other metamorphic rocks

C. Degrees of metamorphism
1. Show in rock's texture and mineralogy
2. Types
   a. Low-grade (e.g. shale becomes slate)
   b. High-grade where the original features are obliterated

D. Metamorphic settings
1. Regional metamorphism
   a. Over extensive areas
   b. Greatest volume of rock produced
2. Contact metamorphism
   a. Near a mass of magma
   b. "Bakes" surrounding rock

E. Metamorphic agents
1. Heat
2. Pressure
   a. From burial
   b. From stress
3. Chemically active fluids
   a. Water (most common)
   b. Ion exchange among minerals

F. Textures
1. Foliated
   a. Minerals in parallel alignment
   b. Minerals perpendicular to force
2. Nonfoliated
   a. Contain equidimensional crystals
   b. Resembles coarse igneous rock

G. Classification
1. Based on texture
2. Two groups
   a. Foliated rocks
      1. Slate
         a. Fine-grained
         b. Splits easily
      2. Schists
         a. Strongly foliated
         b. "Platy"
         c. Types based on composition (e.g., mica schist)
      3. Gneiss
         a. Elongated, granular minerals
         b. Strong segregation of silicates
         c. "Banded" texture
   b. Nonfoliated rocks
      1. Marble
         a. Parent rock—limestone
         b. Calcite crystals
         c. Building stone
         d. Variety of colors
      2. Quartzite
         a. Parent rock—quartz sandstone
         b. Quartz grains are fused

Vocabulary Review

Choosing from the list of key terms, furnish the most appropriate response for the following statements.

1. The ____________ illustrates the origin of the three basic rock types and the role of geologic processes in transforming one rock type into another.
2. The general process of disintegrating and decomposing rock at or near Earth's surface is known as ____________.

Key Terms

Page numbers shown in ( ) refer to the textbook page where the term first appears.

chemical sedimentary rock (p. 42)	lava (p. 32)
chemical weathering (p. 39)	lithification (p. 32)
coarse-grained texture (p. 33)	magma (p. 32)
contact metamorphism (p. 48)	mechanical weathering (p. 38)
crystallization (p. 32)	metamorphic rock (p. 32)
detrital sedimentary rock (p. 42)	nonfoliated (p. 49)
evaporite deposit (p. 45)	porphyritic texture (p. 33)
extrusive (volcanic) (p. 32)	regional metamorphism (p. 48)
exfoliation dome (p. 39)	rock cycle (p. 30)
fine-grained texture (p. 33)	sediment (p. 32)
foliated texture (p. 49)	sedimentary rock (p. 32)
frost wedging (p. 38)	sheeting (p. 39)
glassy texture (p. 34)	strata (beds) (p. 46)
igneous rock (p. 32)	texture (p. 33)
intrusive (plutonic) (p. 32)	weathering (p. 32)

3. ____________ refers to the processes by which sediments are transformed into solid sedimentary rock.
4. Sedimentary rocks, when subjected to great pressures and heat, will turn into the rock type ____________, providing they do not melt.
5. ____________ is molten material found inside Earth.
6. The process whereby magma cools and solidifies, resulting in the formation and growth of a crystalline solid is called ____________.
7. Metamorphic rocks that have formed in response to large-scale mountain building processes are said to have undergone ____________.
8. The rock type that forms when sediment is lithified is ____________.
9. Magma that reaches Earth's surface is called ____________.
10. The term ____________ describes the overall appearance of an igneous rock, based on the size and arrangement of its interlocking crystals.
11. The texture of a metamorphic rock that gives it a layered appearance is called ____________.

12. Igneous rocks that result when lava solidifies are classified as ____________ rocks.
13. Unconsolidated material consisting of particles created by weathering and transported by water, glaciers, wind, or waves is called ____________.
14. The rock type that forms from the crystallization of magma is ____________.
15. Metamorphic rocks composed of only one mineral that forms equidimensional crystals often have a type of texture called a(n) ____________ texture.
16. Igneous rocks that form rapidly at the surface often have a type of texture, called ________, where the individual crystals are too small to be seen with the unaided eye.
17. The mechanical breakup of rock caused by the expansion of freezing water in cracks is called ____________.
18. Detrital sedimentary rocks form from rock fragments, however, a(n) ____________ forms when dissolved substances are precipitated back into solids.
19. An igneous rock that has large crystals embedded in a matrix of smaller crystals is said to have a type of texture called a(n)____________.
20. The single most characteristic feature of sedimentary rocks are layers called ____________.
21. A(n) __________ is a sedimentary deposit that forms as a shallow arm of the sea evaporates.
22. When rock is in contact with, or near, a mass of magma, a type of metamorphism, called ____________, takes place.
23. When igneous rock cools rapidly and ions do not have time to unite into an orderly crystalline structure, a type of texture called a ____________, results.
24. A sedimentary rock that forms from sediments that originate as solid particles from weathered rocks is called a(n)____________.
25. When large masses of magma solidify far below the surface, they form igneous rocks which exhibit a type of texture where the crystals are roughly equal in size and large enough to be identified with the unaided eye, called a(n) ____________.
26. Stone Mountain, Georgia, is an excellent example of a(n) ____________ that formed as weathering caused slabs of material to separate and spall.

Comprehensive Review

1. Using Figure 2.1, label the Earth material or chemical/physical process illustrated at each lettered position in the rock cycle.

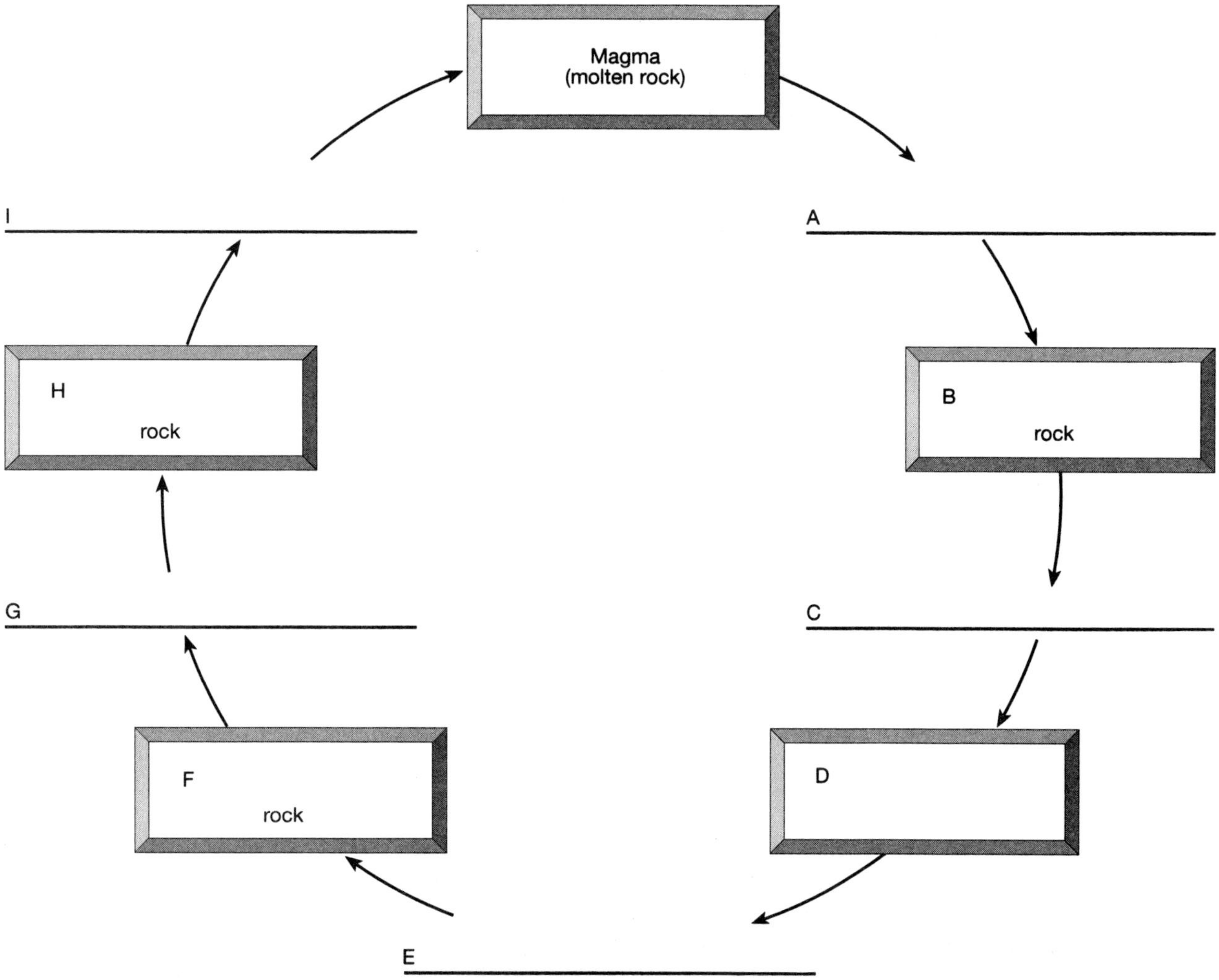

Figure 2.1

2. The rock cycle was initially proposed by ____________ in the late ____________.
3. In the complete rock cycle, any rock type may become ____________ [any other, only one other] rock type.

4. What are the various erosional agents that can pick up, transport, and deposit the products of weathering?

5. Briefly describe the formation of each of the three rock types.

   a) Igneous rocks:

   b) Sedimentary rocks:

   c) Metamorphic rocks:

6. What factor most influences the size of mineral crystals in igneous rocks? How does it affect the size of crystals?

7. What is the difference between magma and lava?

8. List and briefly describe the two criteria used to classify igneous rocks.

   1)

   2)

9. Describe the conditions that will result in the following igneous rock textures. Then, select- the igneous rock photograph in Figure 2.2 that illustrates each of the textures.

   a) Fine-grained texture:

   Photograph letter: ____

   b) Glassy texture:

   Photograph letter: ____

   c) Coarse-grained texture:

   Photograph letter: ____

   d) Porphyritic texture:

   Photograph letter: ____

A.

B.

C.

D.

Figure 2.2

10. What discovery did N.L. Bowen make about the formation of minerals in a cooling magma?

11. According to Bowen's reaction series, the mineral ____________ [quartz, olivine], one of the first minerals to form from magma, will react with the remaining melt and become ____________ [pyroxene, muscovite].

12. Is it possible for two igneous rocks to have the same mineral constituents, but different names? Explain your answer.

13. Define the term *weathering*.

14. List and briefly describe the two kinds of weathering.

    1)

    2)

15. List and briefly describe three natural physical processes that break rocks into smaller fragments.

    1)

    2)

    3)

16. What are the three products of the complete chemical weathering of granite?

17. What are two processes that cause sediment to be lithified into solid sedimentary rock?

18. Describe the two major groups of sedimentary rocks. Give an example of a rock found in each group.

    a) Detrital sedimentary rocks:

       Example:

    b) Chemical sedimentary rocks:

       Example:

19. Detrital sedimentary rocks are subdivided according to ___________ [composition, particle size].

20. What does the presence of angular fragments in a detrital sedimentary rock suggest about the transportation of the particles?

21. What basis is used to subdivide the chemical sedimentary rocks?

22. What is the most abundant chemical sedimentary rock? What are the two origins for this rock?

23. Beginning with a swamp environment, list the successive stages in the formation of coal.

24. Selecting from the sedimentary rock photographs in Figure 2.3, answer the following questions.

    a) The rock in Figure 2.3A is made of __________ [chemical, detrital] material.

    b) The rock in Figure 2.3B is __________ [conglomerate, breccia].

    c) The rock in Figure 2.3 ___ [A, B, C] would most likely be associated with a quiet water environment.

A.

B.

C.

Figure 2.3

25. What are fossils? As important tools for interpreting the geologic past, what are some uses of fossils?

26. Briefly describe the two settings in which metamorphism most often occurs.

    a) Regional metamorphism:

    b) Contact metamorphism:

27. What are the three agents of metamorphism?

28. Describe the two textures of metamorphic rocks. Give an example of a rock that exhibits each texture.

    a) Foliated texture:

       Example:

    b) Nonfoliated texture:

       Example:

29. What is the texture of the metamorphic rock in Figure 2.4?

Figure 2.4

Practice Test

Multiple choice. Choose the best answer for the following multiple choice questions.

1. Which one of the following is NOT an agent of metamorphism?
   a) heat c) chemically active fluids
   b) lithification d) pressure

2. The great majority of rocks exposed at Earth's surface are of which type?
   a) igneous rocks b) sedimentary rocks c) metamorphic rocks

3. Molten material found inside Earth is called ___________.
   a) lava c) plasma e) magma
   b) rock fluid d) mineraloid

4. The greatest volume of metamorphic rock is produced during ___________.
   a) contact metamorphism c) crystallization e) weathering
   b) lithification d) regional metamorphism

5. The rock cycle was initially proposed in the late 1700s by ___________.
   a) James Hutton c) N.L. Bowen e) James Smith
   b) John Playfair d) William Smith

6. Which one of the following is NOT a common cement found in sedimentary rocks?
   a) silica b) gypsum c) calcite d) iron oxide

7. Mechanical weathering by alternate freezing and thawing of water is called ___________.
   a) frost wedging c) sheeting e) exfoliation
   b) unloading d) decomposition

8. Igneous rocks that contain the last minerals to crystallize from magma and consist mainly of feldspars and quartz are said to have a ___________ composition.
   a) granitic c) basaltic e) lithic
   b) gneissic d) metamorphic

9. Fossils found in sedimentary rocks can be used to ___________.
   a) interpret past environments
   b) indicate certain periods of time
   c) match rocks of the same age that are found at different places
   d) all of the above

10. When large masses of magma solidify far below the surface, they form igneous rocks that exhibit a ___________ texture.
    a) fine-grained c) coarse-grained e) fragmental
    b) glassy d) porphyritic

11. Which one of the following is NOT a detrital sedimentary rock?
    a) shale c) sandstone e) siltstone
    b) limestone d) conglomerate

12. The surface process that slowly disintegrates and decomposes rock is called ___________.
    a) erosion c) exfoliation e) crystallization
    b) weathering d) metamorphism

13. Which one of the following is NOT composed of microcrystalline quartz?
    a) travertine
    b) flint
    c) jasper
    d) chert
    e) agate

14. The single most characteristic feature of sedimentary rocks are ___________.
    a) crystals
    b) mud cracks
    c) ripple marks
    d) strata

15. Which one of the following is NOT a primary element found in magma?
    a) silicon
    b) aluminum
    c) iron
    d) carbon
    e) oxygen

16. Metamorphic rocks can form from ___________.
    a) igneous rocks
    b) sedimentary rocks
    c) other metamorphic rocks
    d) all of the above

17. The most important agent of chemical weathering is ___________.
    a) soil
    b) water
    c) carbon
    d) ozone
    e) sulfuric acid

18. Which one of the following metamorphic rocks has a nonfoliated texture?
    a) marble
    b) slate
    c) mica schist
    d) gneiss

19. The term meaning "conversion into rock" is ___________.
    a) metamorphism
    b) crystallization
    c) ionization
    d) lithification
    e) weathering

20. Accumulation of silts and clays, and eventually the rocks siltstone and shale, is generally bestassociated with which one of the following environments?
    a) swiftly flowing river
    b) quiet swamp water
    c) glacier
    d) beach
    e) windy desert

21. Which one of the following is the metamorphic form of coal?
    a) lignite
    b) bituminous
    c) anthracite
    d) peat

22. Which one of the following is NOT a product of the chemical weathering of granite?
    a) clay
    b) carbon dioxide
    c) potassium ions
    d) silica in solution

23. Why can two igneous rocks have the same minerals but different names?
    a) they may have different colors
    b) names of igneous rocks are arbitrary
    c) the rocks may be of different sizes
    d) they may have different textures
    e) they may have been found in different places

24. Detrital sedimentary rocks are subdivided according to ___________.
    a) particle size
    b) color
    c) hardness
    d) the place where they were found
    e) their age

25. Igneous rocks formed from magma that crystallized at depth are called ___________ or intrusive rocks.
    a) plutonic
    b) metamorphic
    c) volcanic
    d) porphyritic
    e) foliated

True/false. For the following true/false questions, if a statement is not completely true, mark it false. For each false statement, change the ***italicized*** *word to correct the statement.*

1. ___ The nonfoliated metamorphic equivalent of limestone is *marble.*

2. ___ The most common extrusive igneous rock is *granite.*

3. ___ Many metallic ore deposits are formed by the precipitation of minerals from *hydrothermal* solutions.

4. ___ The *rock cycle* shows the relations among the three rock types, and is essentially an outline of physical geology.

5. ___ About 75 percent of all rock outcrops on the continents are *sedimentary.*

6. ___ *Granite* is a classic example of an igneous rock that exhibits a coarse-grained texture.

7. ___ *Igneous* rocks are the rock type most likely to contain fossils.

8. ___ Igneous rock, when subjected to heat and pressure far below Earth's surface, will change to *sedimentary* rock.

9. ___ *Mechanical* weathering actually alters what a rock is, changing it into a different substance.

10. ___ Because magma's density is *greater* than the surrounding rocks, it works its way to the surface over time spans from thousands to millions of years.

11. ___ Mineral alignment in a metamorphic rock usually gives the rock a *foliated* texture.

12. ___ Normal rain water is mildly *acidic.*

13. ___ The process, called *weathering,* whereby magma cools, solidifies, and forms igneous-rocks, may take place either beneath the surface, following a volcanic eruption, or at the surface.

14. ___ Separating s*trata* are bedding planes, flat surfaces along which rocks tend to separate or break.

15. ___ *Lithification* literally means to "change form."

16. ___ Molten material found on Earth's surface is called *magma.*

17. ___ Rock salt and rock gypsum form when *evaporation* causes minerals to precipitate from water.

18. ___ The texture of an igneous rock is based in the size and *arrangement* of its interlocking crystals.

19. ___ The two major groups of *metamorphic* rocks are detrital and chemical.

20. ___ The *lithification* of sediment produces sedimentary rock.

21. ___ The agents of *metamorphism* include heat, pressure, and chemically active fluids.

22. ___ According to Bowen's reaction series, quartz is often the *last* mineral to crystallize from a melt.

23. ___ During *mechanical* weathering, oxygen dissolved in water will oxidize some materials.

24. ___ *Sedimentary* rocks are major sources of energy resources, iron, aluminum, and manganese.

25. ___ Slow cooling of magma results in the formation of *small* mineral crystals.

Written questions

1. Referring to the rock cycle, explain why any one rock can be the raw material for another.

2. How is Bowen's reaction series related to the classification of igneous rocks?

3. What are the most common minerals in detrital sedimentary rocks? Why are these minerals so abundant?

4. How does mechanical weathering increase the effectiveness of chemical weathering?

5. In what ways do metamorphic rocks differ from the igneous and sedimentary rocks from which they formed?

3

Landscapes Fashioned by Water

Landscapes Fashioned by Water opens with an examination of the process of mass wasting and its role in the evolution of landscapes. Next, using the water cycle as a model, the exchange of water between the oceans, atmosphere, and continents is discussed. The section on running water includes an examination of the factors that control streamflow and their influence on a stream's ability to erode and transport materials. Erosional and depositional features of narrow and wide-valleys, as well as drainage patterns and stages of valley development, are presented.

After an examination of the importance, occurrence, and movement of groundwater, springs, geysers, wells, and artesian wells are investigated. Following a review of some of the environmental problems associated with groundwater, the chapter ends with a look at the formation and features of caves and karst topography.

Learning Objectives

After reading, studying, and discussing this chapter, you should be able to:

- Explain the process of mass wasting.
- Describe the movement of water through the water cycle.
- Describe the process of streamflow and list the factors that influence a stream's ability to erode and transport materials.
- List and describe the major features produced by stream erosion and deposition.
- Describe the two general types of stream valleys and their major features.
- Distinguish between the different types of drainage patterns.
- Describe how a stream valley and the surrounding landscape can change with time.
- Discuss the occurrence and movement of groundwater.
- Describe springs, geysers, wells, and artesian wells.
- List the major environmental problems associated with groundwater.
- Describe the major features of caves and karst topography.

Chapter Review

- *Mass wasting* is the downslope movement of rock and soil under the direct influence of gravity. Although *gravity* is the controlling force of mass wasting, *water influences mass wasting* by saturating the pore spaces and destroying the cohesion between particles. *Oversteepening* is one factor that triggers mass wasting.

- The *water cycle* describes the continuous interchange of water among the oceans, atmosphere, and continents. Powered by energy from the sun, it is a global system in which the atmosphere provides the link between the oceans and continents. The processes involved in the water cycle include *precipitation, evaporation, infiltration* (the movement of water into rocks or soil through cracks and pore spaces), *runoff* (water that flows over the land, rather than infiltrating into the ground), and *transpiration* (the release of water vapor to the atmosphere by plants).

• The factors that determine a stream's *velocity* are *gradient* (slope of the stream channel), *shape, size* and *roughness* of the channel, and the stream's *discharge* (amount of water passing a given point per unit of time, frequently measured in cubic feet per second). Most often, the gradient and roughness of a stream decrease downstream while width, depth, discharge, and velocity increase.

• The two general types of *base level* (the lowest point to which a stream may erode its channel) are 1) *ultimate base level,* and 2) temporary, or *local base level.* Any change in base level will cause a stream to adjust and establish a new balance. Lowering base level will cause a stream to erode, while raising base level results in deposition of material in the channel.

• The work of a stream includes *erosion* (the incorporation of material), *transportation* (as dissolved load, *suspended load,* and *bed load*), and, whenever a stream's velocity decreases, *deposition.*

• Although many gradations exist, the two general types of stream valleys are 1) *narrow V-shaped valleys* and 2) *wide valleys with flat floors.* Because the dominant activity is downcutting toward base level, narrow valleys often contain *waterfalls* and *rapids.* When a stream has cut its channel closer to base level, its energy is directed from side to side, and erosion produces a flat valley floor, or *floodplain.* Streams that flow upon floodplains often move in sweeping bends called *meanders.* Widespread meandering may result in shorter channel segments, called *cutoffs,* and/or abandoned bends, called *oxbow lakes.*

• Common *drainage patterns* produced by streams include 1) *dendritic,* 2) *radial,* 3) *rectangular,* and 4) *trellis.*

• As a resource, *groundwater* represents the largest reservoir of freshwater that is readily available to humans. Geologically, the dissolving action of groundwater produces *caves* and *sinkholes.* Groundwater is also an equalizer of stream flow.

• Groundwater is that water which occupies the pore spaces in sediment and rock in a zone beneath the surface called the *zone of saturation.* The upper limit of this zone is the *water table.* The *zone of aeration* is above the water table where the soil, sediment, and rock are not saturated. Groundwater generally moves within the zone of saturation. The quantity of water that can be stored depends on the *porosity* (the volume of open spaces) of the material. However, the *permeability* (the ability to transmit a fluid through interconnected pore spaces) of a material is the primary factor controlling the movement of groundwater.

• *Springs* occur whenever the water table intersects the land surface and a natural flow of groundwater results. *Wells,* openings bored into the zone of saturation, withdraw groundwater and create roughly conical depressions in the water table known as *cones of depression. Artesian wells* occur when water rises above the level at which it was initially encountered. Most *caverns* form in limestone at or below the water table when acidic groundwater dissolves rock along lines of weakness, such as joints and bedding planes. *Karst topography* exhibits an irregular terrain punctuated with many depressions, called *sinkholes.*

• Some of the current environmental problems involving groundwater include 1) *overuse* by intense irrigation, 2) *land subsidence* caused by groundwater withdrawal, and 3) *contamination.*

Chapter Outline

- I. Mass wasting
  - A. Gravity is the controlling force
  - B. Oversteepening is a triggering factor
  - C. Helps produce stream valleys
  - D. Types of rapid processes
    - 1. Slump
    - 2. Rockslide
    - 3. Mudflow
    - 4. Earthflow
- II. Water cycle
  - A. Illustrates the circulation of Earth's water supply
  - B. Processes
    - 1. Precipitation
    - 2. Evaporation
    - 3. Infiltration
    - 4. Runoff
    - 5. Transpiration
  - C. Cycle is balanced
- III. Running water
  - A. Streamflow
    - 1. Ability to erode and transport materials depends on velocity
    - 2. Factors that determine velocity
      - a. Gradient, or slope
      - b. Channel characteristics
        - 1. Shape
        - 2. Size
        - 3. Roughness
      - c. Discharge
  - B. Upstream-downstream changes
    - 1. Profile
      - a. Cross-sectional view of a stream
      - b. From head to mouth
    - 2. Factors that increase downstream
      - a. Velocity
      - b. Discharge
      - c. Channel size
    - 3. Factors that decrease downstream
      - a. Gradient, or slope
      - b. Channel roughness
  - C. Base level
    - 1. Lowest point a stream can erode to
    - 2. Two general types
      - a. Ultimate
      - b. Temporary, or local
    - 3. Changing causes readjustment of stream
  - D. Work of streams
    - 1. Erosion
    - 2. Transportation (material is called load)
      - a. Types of load
        - 1. Dissolved load
        - 2. Suspended load
        - 3. Bed load
      - b. Load is related to a stream's
        - 1. Competence—maximum size of particles
        - 2. Capacity—maximum load
    - 3. Deposition
      - a. Caused by a decrease in velocity
        - 1. Competence is reduced
        - 2. Sediment begins to drop out
      - b. Stream sediments are called alluvium
      - c. Features produced by deposition
        - 1. Deltas
          - a. Exist in oceans or lakes
          - b. Distributaries often form in channel
        - 2. Natural levees
          - a. Parallel to stream channels
          - b. Area behind may contain
            - 1. Back swamps
            - 2. Yazoo tributaries
  - E. Stream valleys
    - 1. Sides shaped by
      - a. Weathering
      - b. Overland flow
      - c. Mass wasting
    - 2. Narrow valleys
      - a. V-shaped
      - b. Downcutting toward base level
      - c. Characteristic features of
        - 1. Rapids
        - 2. Waterfalls
    - 3. Wide valleys
      - a. Stream near base level
      - b. Downward erosion less dominant
      - c. Energy directed from side to side
      - d. Floodplain
      - e. Characteristic features of
        - 1. Meanders
        - 2. Cutoffs
        - 3. Oxbow lakes
  - F. Drainage basins and patterns
    - 1. A divide separates basins
    - 2. Types of drainage patterns

a. Dendritic
b. Radial
c. Rectangular
d. Trellis
G. Stages of valley development
1. Youth
a. Rapids and waterfalls
b. V-shaped valleys
c. Vigorous downcutting
d. Steep gradient
2. Maturity
a. Downward erosion diminishes
b. Lateral erosion dominates
c. Floodplain beginning
d. Gradient lower than at youth
3. Old age
a. Large floodplain
b. Widespread shifting of stream
c. Characteristic features of
1. Natural levees
2. Backswamps
3. Yazoo tributaries
4. Rejuvenation
a. "Made young again"
b. Characteristic features of
1. Entrenched meanders
2. Terraces
IV. Water beneath the surface (groundwater)
A. Largest freshwater reservoir for humans
B. Geological roles
1. Erosional agent (dissolving)
a. Sinkholes
b. Caverns
2. Equalizer of streamflow
C. Distribution and movement of groundwater
1. Distribution
a. Zone of saturation—all pore spaces filled with water
b. Water table—upper limit of zone of saturation
c. Zone of aeration
1. Area above water table
2. Pore spaces filled with air
2. Movement
a. Porosity—percent of pore spaces
b. Permeability
1. Ability to transmit water
2. Aquitarde—impermeable layer
3. Aquifer—permeable layer
D. Water features
1. Springs
2. Hot springs and geysers
3. Wells
a. Drawdown
b. Cone of depression
4. Artesian wells
a. Water rises higher than initial level
b. Types
1. Nonflowing
2. Flowing
E. Environmental problems associated with groundwater
1. Treating it as a nonrenewable resource
2. Land subsidence caused by its withdrawal
3. Contamination
F. Features produced by groundwater
1. Groundwater dissolves soluble rock because it contains weak carbonic acid
2. Caverns
a. Formed by dissolving rock beneath surface
b. Features formed from dripstone
1. Stalactites
2. Stalagmites
3. Karst topography
a. Formed by dissolving rock at, or near, surface
b. Common features
1. Sinkholes
a. Surface depressions
b. Formed by
1. Dissolving bedrock, or
2. Cavern collapse
c. Lack of good surface drainage

Vocabulary Review

Choosing from the list of key terms, furnish the most appropriate response for the following statements.

1. ____________ is the downslope movement of rock and soil under the direct influence of gravity.
2. The land area that contributes water to a stream is referred to as the stream's ____________.
3. The zone beneath Earth's surface where all open spaces in sediment and rock are completely filled with water is the ____________.
4. The ____________ of a material is the percentage of the total volume that consists of pore spaces.
5. ____________ is the slope of a stream channel expressed as the vertical drop of a stream over a specified distance.
6. The general term for any well sorted stream-deposited sediment is ____________.
7. The ____________of a stream is the maximum load it can carry.
8. Water beneath Earth's surface, in the zone of saturation, is known as ____________.
9. A tributary that flows parallel to the main stream because a natural levee is present iscalled a ____________.
10. The release of water vapor to the atmosphere by plants is referred to as ____________.
11. A "looplike," sweeping bend in a stream is called a ____________.
12. The ____________ is the upper limit of the zone of saturation.
13. The volume of water flowing past a certain point per unit of time is called a stream's ____________.
14. The term ____________ is applied to any well in which groundwater rises above the level where it was initially encountered.
15. The elevated landform that parallels some streams, called a ____________, acts to confine the stream's waters, except during flood stage.
16. The ____________ of a material indicates its ability to transmit water through interconnected pore spaces.

Key Terms

Page numbers shown in ( ) refer to the textbook page where the term first appears.

alluvium (p. 66)
aquitard (p. 77)
aquifer (p. 77)
artesian well (p. 79)
backswamp (p. 67)
base level (p. 64)
bed load (p. 64)
capacity (p. 65)
cavern (p. 82)
competence (p. 65)
cone of depression (p. 78)
cutoff (p. 70)
delta (p. 66)
dendritic pattern (p. 70)
discharge (p. 63)
dissolved load (p. 64)
distributary (p. 66)
divide (p. 70)
drainage basin (p. 70)
drawdown (p.78)
entrenched meander (p. 74)
floodplain (p. 69)
geyser (p. 78)
gradient (p. 62)
groundwater (p. 75)
hot spring (p. 78)
karst topography (p. 83)
mass wasting (p. 58)
meander (p. 69)
natural levee (p. 67)
oxbow lake (p. 70)
permeability (p. 77)
porosity (p. 77)
radial pattern (p. 71)
rectangular pattern (p. 71)
rejuvenation (p. 74)
sinkhole (sink) (p. 84)
sorting (p. 66)
spring (p. 77)
stalactite (p. 83)
stalagmite (p. 83)
suspended load (p. 64)
transpiration (p. 61)
trellis pattern (p. 71)
water cycle (p. 60)
water table (p. 75)
well (p. 78)
yazoo tributary (p. 67)
zone of aeration (p. 75)
zone of saturation (p. 75)

17. The unending circulation of Earth's water supply from the oceans, to the atmosphere, to the land and back to the oceans is called the ____________.

18. A(n) ____________ is the flat, low-lying portion of a stream valley that is subject to inundation during flooding.

19. The ____________ of a stream measures the maximum size of particles it is capable of transporting.

20. The drainage basin of one stream is separated from the drainage basin of another by an imaginary line called a(n) ____________.

21. An impermeable layer that hinders or prevents the movement of groundwater is termed a(n) ____________.

22. ____________ is the lowest point to which a stream can erode its channel.

23. Whenever the water table intersects the ground surface, a natural flow of groundwater, called a ____________, results.

24. A rock strata or sediment that transmits groundwater freely is called a(n) ____________.

25. The fine sediment transported within the body of flowing water is called a stream's ____________.

26. Landscapes that, to a large extent, have been shaped by the dissolving power of groundwater are said to exhibit ____________.

Comprehensive Review

1. Define mass wasting and list the factors that play an important part in the process.

2. Identify each of the forms of mass wasting illustrated in Figure 3.1 by writing the name of the process below the diagram.

3. The combined effects of which two processes produce stream valleys?

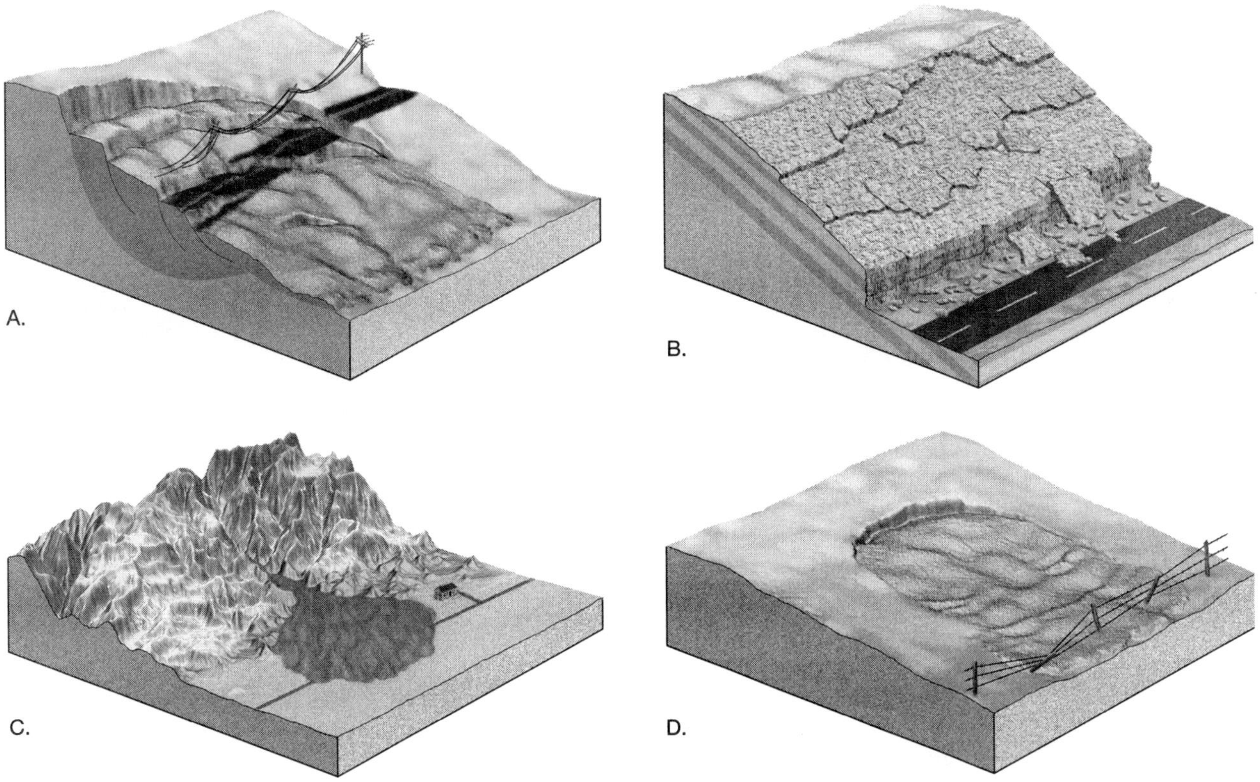

Figure 3.1

4. Figure 3.2 illustrates the water cycle. Select the letter in the figure that represents each of the following:

   a) Runoff: ____ b) Evaporation: ____ c) Precipitation: ____ d) Infiltration: ____

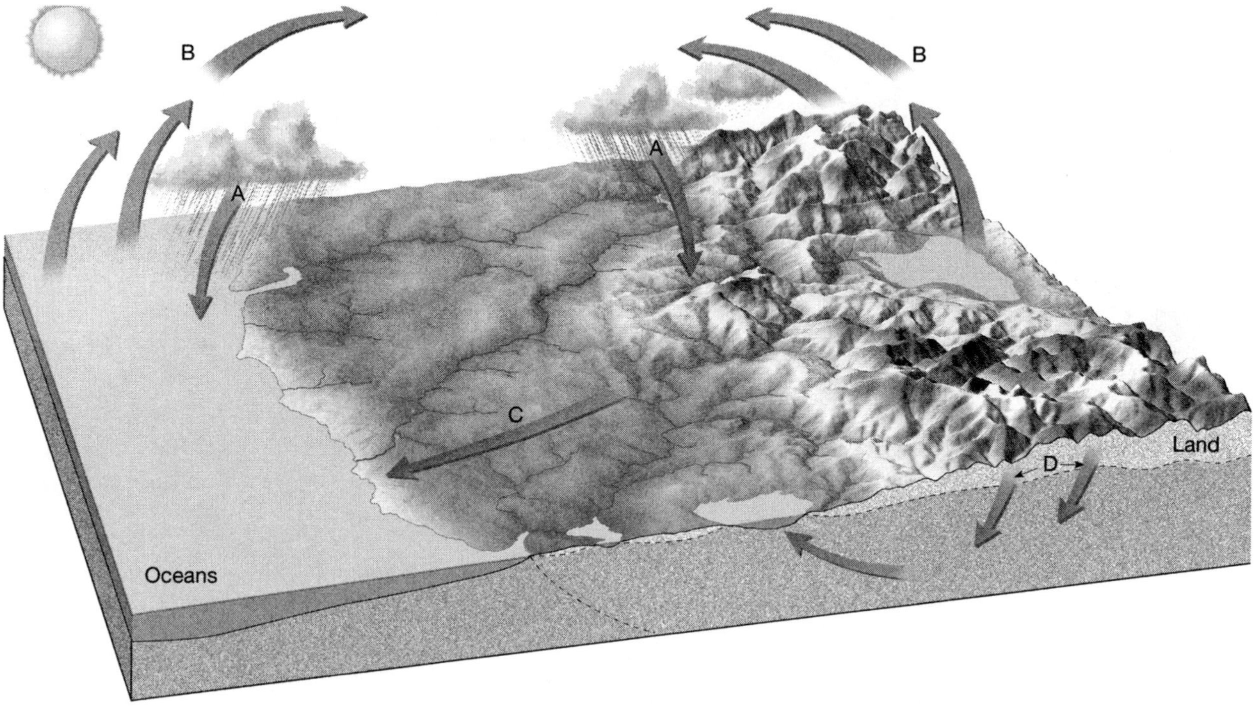

Figure 3.2

5. Briefly define the following measurements that are used to assess streamflow:
   a) Velocity:

   b) Gradient:

   c) Discharge:

6. What is the average gradient of a stream whose headwaters are 10 kilometers from the sea, at an altitude of 40 meters?

7. By circling the correct response, indicate how each of the following measurements changes from a stream's head to its mouth.
   a) Gradient [increases, decreases]d) Discharge [increases, decreases]
   b) Velocity [increases, decreases]e) Width and depth [increase, decrease]
   c) Channel roughness [increases, decreases]

8. Describe the two general types of base level.
   a) Ultimate base level:

   b) Temporary, or local, base level:

9. List and describe the three ways streams transport their load of sediment.

   1)

   2)

   3)

10. Why does the greatest erosion and transportation of sediment by a stream occur during a flood?

11. Figure 3.3 illustrates a wide stream valley. Select the letter in the figure that identifies each of the following features:

    a) Yazoo tributary: ____ c) Natural levees: ____

    b) Back swamp: ____ d) Oxbow lake: ____

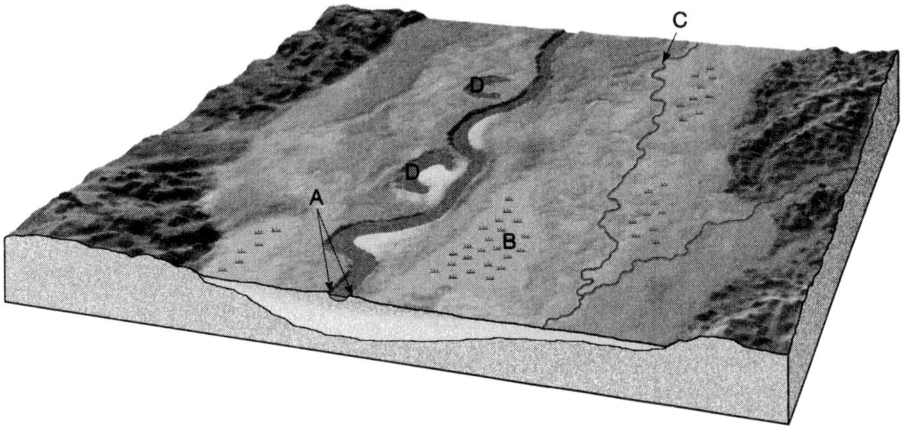

Figure 3.3

12. Briefly describe the appearance of the following types of valleys. Also list the major features you would expect to find associated with each type.

    1) Narrow valley:

    Features:

    2) Wide valley:

    Features:

13. Describe the type of drainage pattern that would develop on an isolated volcanic cone or domal uplift.

14. From the illustrations in Figure 3.4, select the diagram that best illustrates each of the following stages in the evolution of a valley.

    a) Youth: ____ b) Maturity: ____ c) Old age: ____

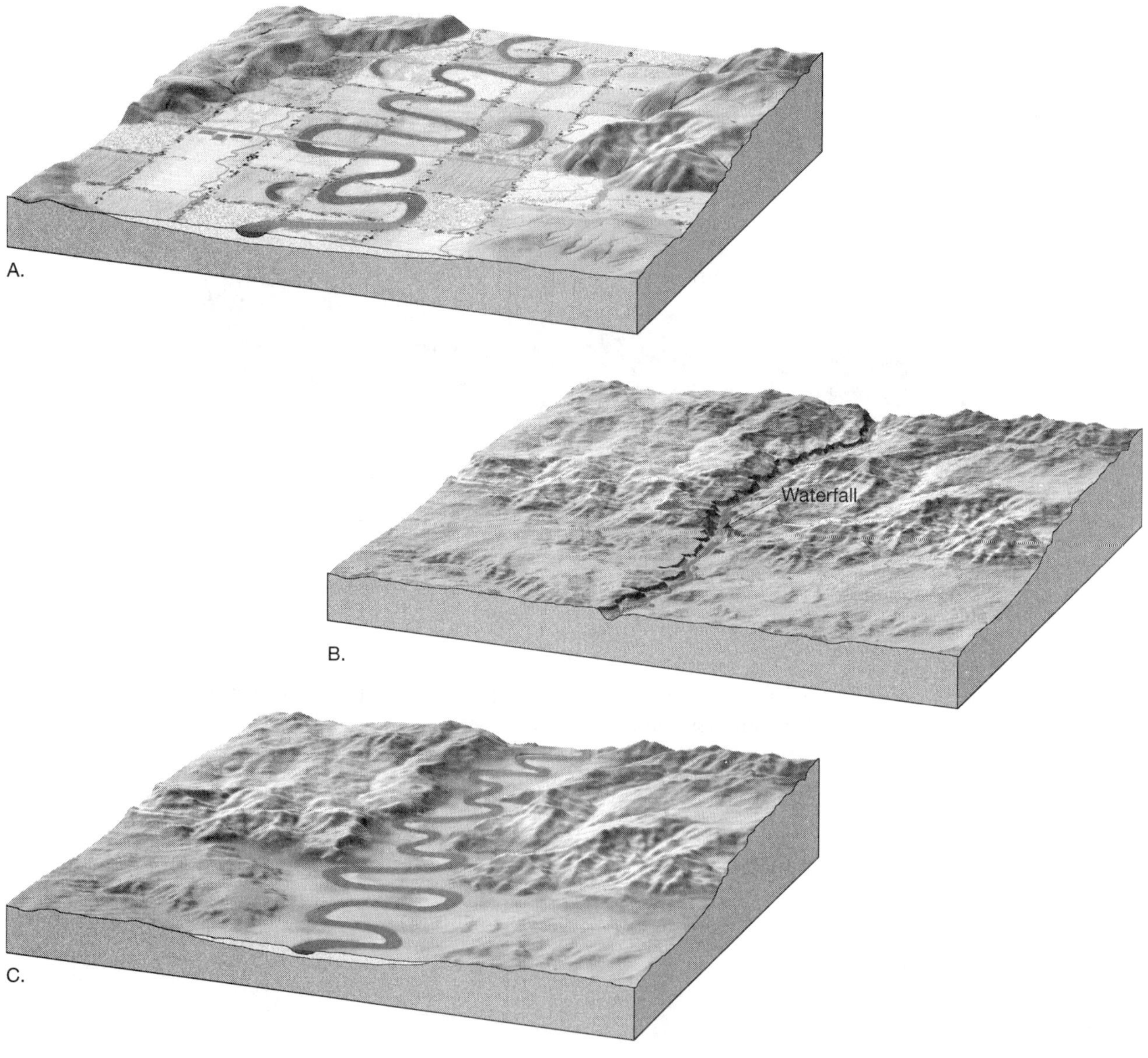

Figure 3.4

15. List some key words or features that describe or are associated with each of the three stages of valley development.

    a) Youth:

    b) Mature:

    c) Old age:

16. Using Figure 3.5, select the letter that illustrates each of the following:

    a) Zone of saturation: ____

    b) Aquitard: ____

    c) Spring: ____

    d) Water table: ____

    e) Zone of aeration: ____

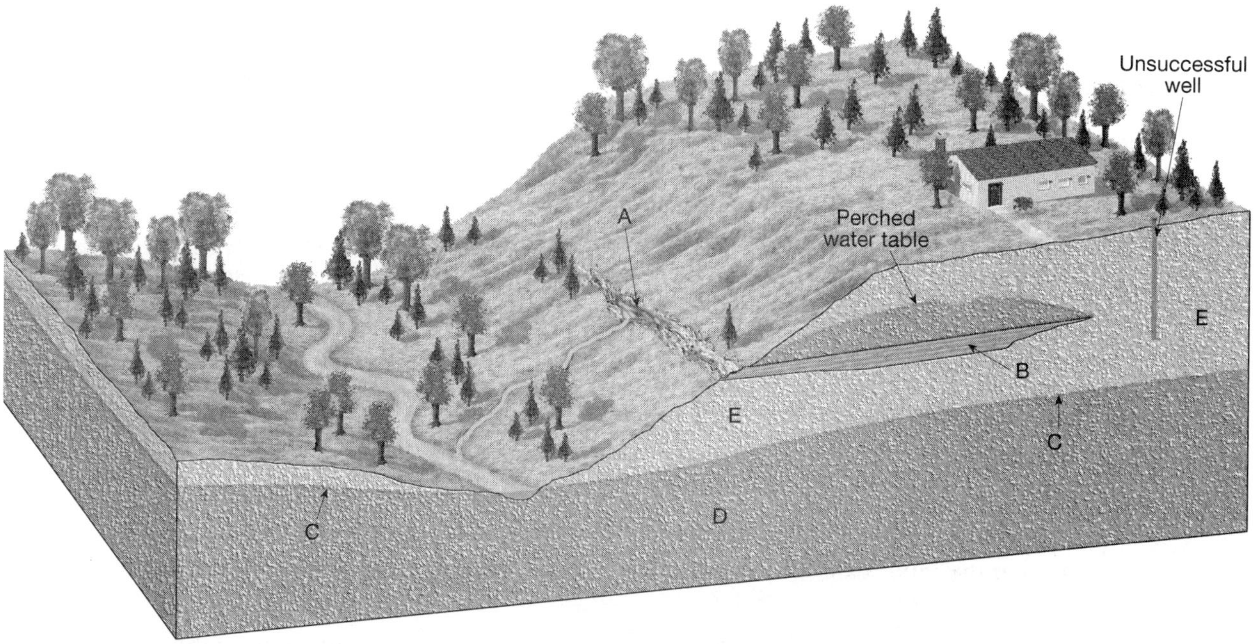

Figure 3.5

17. In general, how is the shape of the water table related to the shape of the surface?

18. What is the source of heat for most hot springs and geysers?

19. Using the terms *porosity* and *permeability*, describe an aquifer.

20. List two environmental problems that are associated with groundwater.

    1)

    2)

21. What are two ways that sinkholes form?

    1)

    2)

Practice Test

Multiple choice. Choose the best answer for the following multiple choice questions.

1. The largest reservoir of freshwater that is available to humans is ____________.
   a) lakes and reservoirs c) groundwater e) atmospheric water vapor
   b) glaciers d) river water

2. The single most important agent sculpturing Earth's land surface is ____________.
   a) ice b) wind c) running water d) waves

3. Lowering a stream's base level will cause the stream to ____________.
   a) deposit c) change course e) stop flowing
   b) meander d) erode

4. Along a stream meander, the maximum velocity of water occurs ____________.
   a) near the inner bank of the meander c) near the outer bank of the meander
   b) in the center d) near the stream's bed

5. A typical profile of a stream exhibits a constantly __________ gradient from the head to the mouth.
   a) increasing b) decreasing c) uniform d) arbitrary

6. The downslope movement of material under the direct influence of gravity is called ___________.
   a) erosion   c) competence   e)talus
   b) mass wasting   d) weathering

7. Which type of drainage pattern will most likely develop where the underlying material is relatively uniform?
   a) dendritic   b) radial   c) rectangular   d) trellis

8. Ice sheets and glaciers represent approximately ___________ percent of the total volume of freshwater.
   a) 15   b) 35   c) 55   d) 85

9. The function of artificial levees built along a river is to control ___________.
   a) erosion   c) gradient   e) oxbow lakes
   b) flooding   d) meandering

10. Loose, undisturbed particles assume a stable slope called the angle of ___________.
    a) slope   c) incidence   e) slump
    b) stability   d) repose

11. Occasionally, deposition causes the main channel of a stream to divide into several smaller channels called ___________.
    a) deltas   c) oxbows   e) distributaries
    b) meanders   d) yazoos

12. The average annual precipitation worldwide must equal the quantity of water ___________.
    a) evaporated   c) that becomes runoff   e) locked in glaciers
    b) transpired   d) infiltrated

13. During periods when rain does not fall, rivers are sustained by ___________.
    a) transpirated water   b) groundwater
    c) atmospheric moisture   d) runoff

14. Which one of the following would least likely be found in a wide valley?
    a) oxbow lake   c) rapids   e) meanders
    b) floodplain   d) cutoff

15. Along straight stretches, the highest velocities of water are near the ___________ of the channel, just below the surface.
    a) banks   b) center

16. The combined effects of ___________ and running water produce stream valleys.
    a) weathering   c) mass wasting   e) capacity
    b) tectonics   d) discharge

17. Which one of the following is a measure of a material's ability to transmit water through interconnected pore spaces?
    a) competence   c) capacity   e) permeability
    b) gradient   d) porosity

18. The capacity of a stream is directly related to its ___________.
    a) velocity   c) gradient   e) discharge
    b) competence   d) meandering

19. Drawdown of a well creates a roughly conical lowering of the water table called the code of______.

a) saturation c) permeability e) aeration
b) lowering d) depression

20. In a typical stream, where gradient is steep, discharge is ____________.

a) large c) variable e) reversed
b) small d) impossible to calculate

21. Most natural water contains the weak acid, ____________ acid, which makes it possible for groundwater to dissolve limestone and form caverns.

a) acetic c) carbonic e) nitric
b) hydrochloric d) sulfuric

22. The area above the water table where soil is not saturated is called the zone of ___________.

a) aeration c) saturation e) aquifers
b) soil moisture d) porosity

23. Most streams carry the largest part of their load ____________.

a) as bed load b) in suspension c) in solution d) near their head

24. The great majority of hot springs and geysers in the United States are found in the ____________.

a) North c) East e) Midwest
b) South d) West

25. Whenever the water table intersects the ground surface, a natural flow of water, called a ____________, results.

a) spring c) aquiclude e) geyser
b) stalagmite d) sinkhole

True/false. For the following true/false questions, if a statement is not completely true, mark it false. For each false statement, change the ***italicized*** *word to correct the statement.*

1. ___ Most of a stream's dissolved load is brought to it by *groundwater.*

2. ___ *Ultimate* base levels include lakes and resistant layers of rock.

3. ___ During mass wasting, water in the pore spaces of sediment *destroys* the cohesion between particles, which aids the downslope movement of material.

4. ___ The reduced current velocity at the inside of a meander results in the deposition of coarse sediment, especially sand, called a *point bar.*

5. ___ Landscapes shaped by the dissolving power of groundwater are often said to exhibit *karst* topography.

6. ___ A flat valley floor, or floodplain, often occurs in *narrow* valleys.

7. ___ Since the water cycle is balanced, the average annual *precipitation* worldwide must equal the quantity of water evaporated.

8. ___ The gradient of a stream *increases* downstream.

9. ___ The drainage pattern that develops when the bedrock is crisscrossed by many right-angle joints and/or faults is the *radial* pattern.

10. ___ The *highest* water velocities in a straight channel are near the center, just below the surface.

11. ___ Much of the water that flows in rivers is not direct runoff from rain or snowmelt, but originates as *groundwater.*

12. ___ *Permeable* layers, such as clay, that hinder or prevent the movement of water beneath the surface are termed aquitards

13. ___ The term *artesian* is applied to any situation in which groundwater rises in a well above the level where it was initially encountered.

14. ___ A channel's shape, size, and *roughness* affect the amount of friction with the water in the channel.

15. ___ When rivers become rejuvenated, meanders often stay in the same place but become *deeper.*

16. ___ The downslope movement of rock and soil under the direct influence of *ice* is called mass wasting.

17. ___ The *competence* of a stream measures the maximum size of particles it is capable of transporting.

18. ___ A stream's velocity *increases* downstream.

19. ___ *Lowering* base level will cause a stream to gain energy and erode.

20. ___ Marshes called *backswamps* form because water on a floodplain cannot flow up the levee and into the river.

21. ___ Mass wasting is the step that follows *weathering* in the evolution of most landscapes.

22. ___ *Groundwater* represents the largest reservoir of freshwater that is readily available to humans.

23. ___ Some of the water that infiltrates the ground surface is absorbed by plants, which then release it into the atmosphere through a process called *infiltration.*

24. ___ Sorting of sediment occurs because each particle *size* has a critical settling velocity.

25. ___ Near the *mouth* of a river where discharge is great, gradient is small.

26. ___ The zone in the ground where all of the open spaces in sediment and rock are completely filled with water is called the zone of *saturation.*

27. ___ *Geysers* are intermittent hot springs that eject water with great force at various intervals.

28. ___ Two types of floodplains are erosional floodplains and *depositional* floodplains.

29. ___ Land *subsidence* is an environmental problem that can be caused by groundwater withdrawal.

30. ___ *Ice* is the single most important agent sculpturing Earth's land surface.

Written questions

1. Briefly describe the movement of water through the water cycle.

2. List and briefly describe the three stages of valley development.

3. What is the difference between the porosity of a material and its permeability?

4. What are the conditions required for the development of karst topography?

Glacial and Arid Landscapes

4

Glacial and Arid Landscapes begins by examining the types, the movement, and the formation of glaciers. Also presented are the processes of glacial erosion and deposition, as well as the features associated with valley glaciers and ice sheets. The Pleistocene epoch and some indirect effects of Ice Age glaciers are also discussed.

The climatic conditions that produce the semiarid and arid regions opens the discussion of the world's dry regions. The role of water in arid climates is examined. A discussion of the evolution of much of the Basin and Range region of the United States provides insight into the processes that shape desert landscapes. The chapter ends with an investigation of wind erosion and deposition.

Learning Objectives

After reading, studying, and discussing this chapter, you should be able to:

- Describe the types and locations of glaciers.
- Discuss glacial movement.
- List the types of glacial drift.
- Describe the features produced by glacial erosion and deposition.
- Define the Ice Age and the Pleistocene epoch.
- Describe the climatic conditions that produce arid and semiarid regions.
- Describe the geologic evolution of the Basin and Range region
- List the types of wind deposits and describe the features of wind deposition.

Chapter Review

- A *glacier* is a thick mass of ice originating on the land as a result of the compaction and recrystallization of snow, and shows evidence of past or present flow. Today, *valley* or *alpine glaciers* are found in mountain areas where they usually follow valleys that were originally occupied by streams. *Ice sheets* exist on a much larger scale, covering most of Greenland and Antarctica.

- On the surface of a glacier, ice is brittle. However, below about 50 meters, pressure is great, causing ice to *flow* like a *plastic material.* A second important mechanism of glacial movement consists of the whole ice mass *slipping* along the ground.

- Glaciers erode land by *plucking* (lifting pieces of bedrock out of place) and *abrasion* (grinding and scraping of a rock surface). Erosional features produced by valley glaciers include *glacial troughs, hanging valleys, cirques, arêtes, horns, and fiords.*

• Any sediment of glacial origin is called *drift.* The two distinct types of glacial drift are 1) *till,* which is material deposited directly by the ice; and 2) *stratified drift,* which is sediment laid down by meltwater from a glacier.

• The most widespread features created by glacial deposition are layers or ridges of till, called *moraines.* Associated with valley glaciers are *lateral moraines,* formed along the sides of the valley, and *medial moraines,* formed between two valley glaciers that have joined. *End moraines,* which mark the former position of the front of a glacier, and *ground moraine,* an undulating layer of till deposited as the ice front retreats, are common to both valley glaciers and ice sheets.

• Perhaps the most convincing evidence for the occurrence of several glacial advances during the *Ice Age* is the widespread existence of *multiple layers of drift* and an uninterrupted record of climate cycles preserved in *sea floor sediments.* In addition to massive erosional and depositional work, other effects of Ice Age glaciers included the *forced migration* of animals, *changes in stream and river courses, adjustment of the crust* by rebounding after the removal of the immense load of ice, and *climate changes* caused by the existence of the glaciers themselves. In the sea, the most far-reaching effect of the Ice Age was the *worldwide change in sea level* that accompanied each advance and retreat of the ice sheets.

• Deserts in the lower latitudes coincide with zones of high air pressure known as *subtropical highs.* Middle latitude deserts exist because of their positions in the deep interiors of large continents, far removed from oceans. Mountains also act to shield these regions from marine air masses.

• Practically all desert streams are dry most of the time and are said to be *ephemeral.* Nevertheless, *running water is responsible for most of the erosional work in a desert.* Although wind erosion is more significant in dry areas than elsewhere, the main role of wind in a desert is in the transportation and deposition of sediment.

• Many of the landscapes of the Basin and Range region of the western and southwestern United States are the result of streams eroding uplifted mountain blocks and depositing the sediment in interior basins. *Alluvial fans, playas,* and *playa lakes* are features often associated with these landscapes.

• In order for wind erosion to be effective, dryness and scant vegetation are essential. *Deflation,* the lifting and removal of loose material, often produces shallow depressions called *blowouts* and can also lower the surface by removing sand and silt, leaving behind a stony veneer called *desert pavement. Abrasion,* the "sandblasting" effect of wind, is often given too much credit for producing desert features. However, abrasion does cut and polish rock near the surface.

• Wind deposits are of two distinct types: 1) extensive *blankets of silt,* called *loess,* that is carried by wind in *suspension;* and 2) *mounds and ridges of sand,* called *dunes,* which are formed from sediment that is carried as part of the wind's bed load.

Chapter Outline

I. Glaciers
  A. Types
    1. Valley, or alpine
    2. Ice sheets, or continental
  B. Movement of glacial ice
    1. Types
      a. Plastic movement
        1. Occurs below 50 meters
        2. Zone of fracture
          a. Uppermost 50 meters
          b. Crevasses form
      b. Slipping along the ground
    2. Zone of accumulation—where ice forms
    3. Zone of wastage—where ice melts

- C. Glaciers erode by
  - 1. Plucking—lifting of rock blocks
  - 2. Abrasion
    - a. Rock flour—pulverized rock
    - b. Striations—grooves on bedrock
- D. Erosional features of valley glaciers
  - 1. Glacial trough
  - 2. Hanging valley
  - 3. Cirque
  - 4. Arête
  - 5. Horn
  - 6. Fiord
- E. Glacial deposits
  - 1. Glacial drift
    - a. All sediments of glacial origin
    - b. Types
      - 1. Till
        - a. Deposited directly by ice
        - b. Glacial erratics—boulders in till
      - 2. Stratified drift
        - a. Deposited by meltwater
        - b. Sediment is sorted
  - 2. Depositional features
    - a. Moraines
      - 1. Layers or ridges of till
      - 2. Types
        - a. Lateral
        - b. Medial
        - c. End
        - d. Ground
    - b. Outwash plain, or valley train
    - c. Kettles
    - d. Drumlins
    - e. Eskers
    - f. Kames
- F. Glaciers of the past
  - 1. Ice Age
    - a. Began 2 to 3 million years ago
    - b. Ice covered 30% of land area
    - c. Division of time called pleistocene epoch
  - 2. Indirect effects
    - a. Migration of animals and plants
    - b. Rebounding upward of crust
    - c. Worldwide change in sea level
    - d. Climatic changes

II. Deserts

- A. Types of dry lands
  - 1. Desert, or arid
  - 2. Steppe, or semiarid
    - a. More humid than desert
    - b. Transition zone surrounding desert
- B. Distribution and causes
  - 1. Dry lands at lower latitudes are.caused by high air pressure
  - 2. Dry lands at middle latitudes are.caused by position far removed from oceans
- C. Role of water
  - 1. Streams are dry most of the time
    - a. Flow only during periods of rainfall (ephemeral streams)
    - b. Types
      - 1. Wash
      - 2. Arroyo
      - 3. Wadi, donga, or nullah
  - 2. Desert rainfall
    - a. Often occurs as heavy showers
    - b. Causes flash floods
  - 3. Poorly integrated drainage
  - 4. Most erosional work is done by water
- D. Evolution of a desert landscape
  - 1. Uplifted crustal blocks
  - 2. Interior drainage produces
    - a. Alluvial fans
    - b. Playas, playa lakes
- E. Wind erosion
  - 1. By deflation
    - a. Lifting of loose material
    - b. Produces
      - 1. Blowouts
      - 2. Desert pavement
  - 2. By abrasion
- F. Types of wind deposits
  - 1. Loess
    - a. Deposits of windblown silt
    - b. Extensive blanket deposits
    - c. Primary sources are
      - 1. Deserts
      - 2. Glacial stratified drift
  - 2. Sand dunes
    - a. Mounds and ridges of sand from wind's bed load
    - b. Characteristic features
      - 1. Slip face—leeward slope
      - 2. Cross beds contain sloping layers

Vocabulary Review

Choosing from the list of key terms, furnish the most appropriate response for the following statements.

1. The largest type of glacier is the ___________, which often covers a large portion of a continent.
2. A(n) ___________ is type of stream that carries water only in response to a specific episode of rainfall.
3. The all-embracing term for sediments of glacial origin, no matter how, where, or in what form they were deposited is ___________.
4. A(n) ___________ is a streamlined asymmetrical hill composed of till.
5. Most of the major glacial episodes occurred during a division of the geologic time scale known as the ___________.
6. Embedded material in a glacier may gouge long scratches and grooves called ___________ in the bedrock as the ice passes over.
7. Arid regions typically lack permanent streams and are often characterized by a type of drainage called ___________.
8. A(n) ___________ is a type of moraine that forms when two valley glaciers coalesce to form a single ice stream.
9. The area where there is net loss to a glacier due to melting is known as the ___________.
10. A(n) ___________ is a U-shaped valley produced by theerosion of a valley glacier.
11. A(n) ___________ is a hollowed out, bowl-shaped depression at the head of a valley produced by glacial erosion.
12. A mound or ridge of wind-deposited sand is a(n) ___________.
13. The part of a glacier where snow accumulates and ice forms is called the ___________.
14. As a stream emerges from a canyon and quickly loses velocity, it often deposits a cone of debris known as a(n) ___________.
15. The climatic type known as a(n)___________ is semiarid and often surrounds a desert.
16. Deposits of windblown silt are called ___________.

Key Terms

Page numbers shown in ( ) refer to the textbook page where the term first appears.

alluvial fan (p. 106)
alpine glacier (p. 90)
arête (p. 94)
blowout (p. 108)
cirque (p. 94)
crevasse (p. 91)
cross beds (p. 110)
deflation (p. 108)
desert (p. 103)
desert pavement (p. 108)
drift (p. 96)
drumlin (p. 99)
dune (p. 110)
end moraine (p. 98)
ephemeral stream (p. 103)
esker (p. 99)
fiord (p. 95)
glacial erratic (p. 98)
glacial striations (p. 93)
glacial trough (p. 94)
glacier (p. 90)
ground moraine (p. 99)
hanging valley (p. 94)
horn (p. 94)
ice sheet (p. 91)
interior drainage (p. 105)
kame (p. 100)
kettle (p. 99)
lateral moraine (p. 98)
loess (p. 109)
medial moraine (p. 98)
outwash plain (p. 99)
playa lake (p. 106)
Pleistocene epoch (p. 101)
plucking (p. 93)
pluvial lake (p. 102)
rock flour (p. 93)
slip face (p. 110)
steppe (p. 103)
stratified drift (p. 98)
till (p. 97)
valley glacier (p. 90)
valley train (p. 99)
zone of accumulation (p. 91)
zone of wastage (p. 91)

17. Materials deposited directly by glacial ice are known as ___________.

18. A broad ramplike surface of stratified drift that is built adjacent to the downstream edge of an ice sheet is termed a(n) ___________.

19. Deflation by wind often forms a shallow depression called a(n) ___________.

20. The term applied to the sloping layers of sand that compose a dune is ___________.

21. A(n) ___________ is a thick mass of ice that forms over land from the accumulation, compaction and recrystallization of snow.

Comprehensive Review

1. List and describe the two major types of glaciers.

   1)

   2)

2. Why are the uppermost 50 meters of a glacier appropriately referred to as the zone of fracture?

3. Briefly describe the following two ways that glaciers erode land.

   a) Plucking:

   b) Abrasion:

4. The area represented in Figure 4.1 was subjected to alpine glaciation. Select the appropriate letter in the figure that identifies each of the following features.

   a) Cirque: ____

   b) Glacial trough: ____

   c) Hanging valley: ____

   d) Horn: ____

   e) arête: ____

5. Distinguish between the two terms *till* and *stratified drift.*

6. Describe the difference between the two dry climatic types, desert and steppe.

   a) Desert:

   b) Steppe:

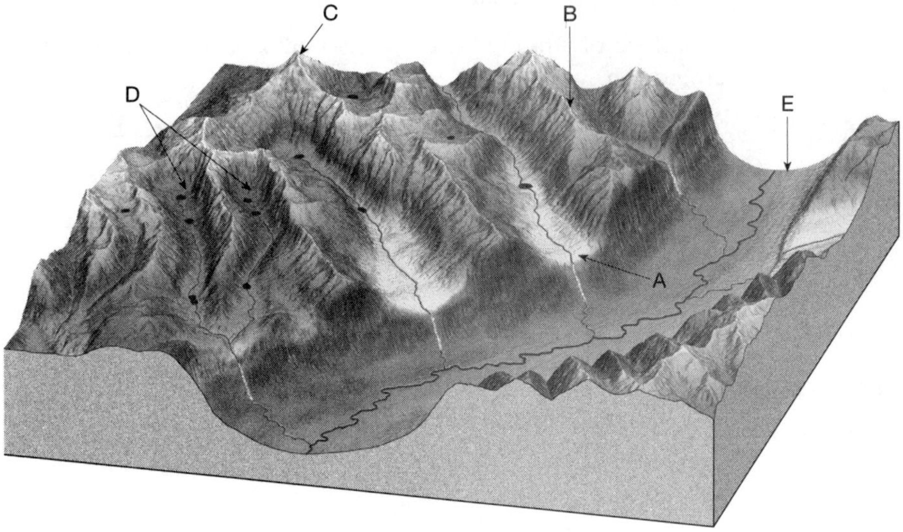

Figure 4.1

7. In Figure 4.2, diagram ____ [A, B] illustrates the early stage in the evolution of the desert landscape. Select the appropriate letter in the figure that identifies each of the following features.

   a) Playa lake: ____ b) Alluvial fan: ____

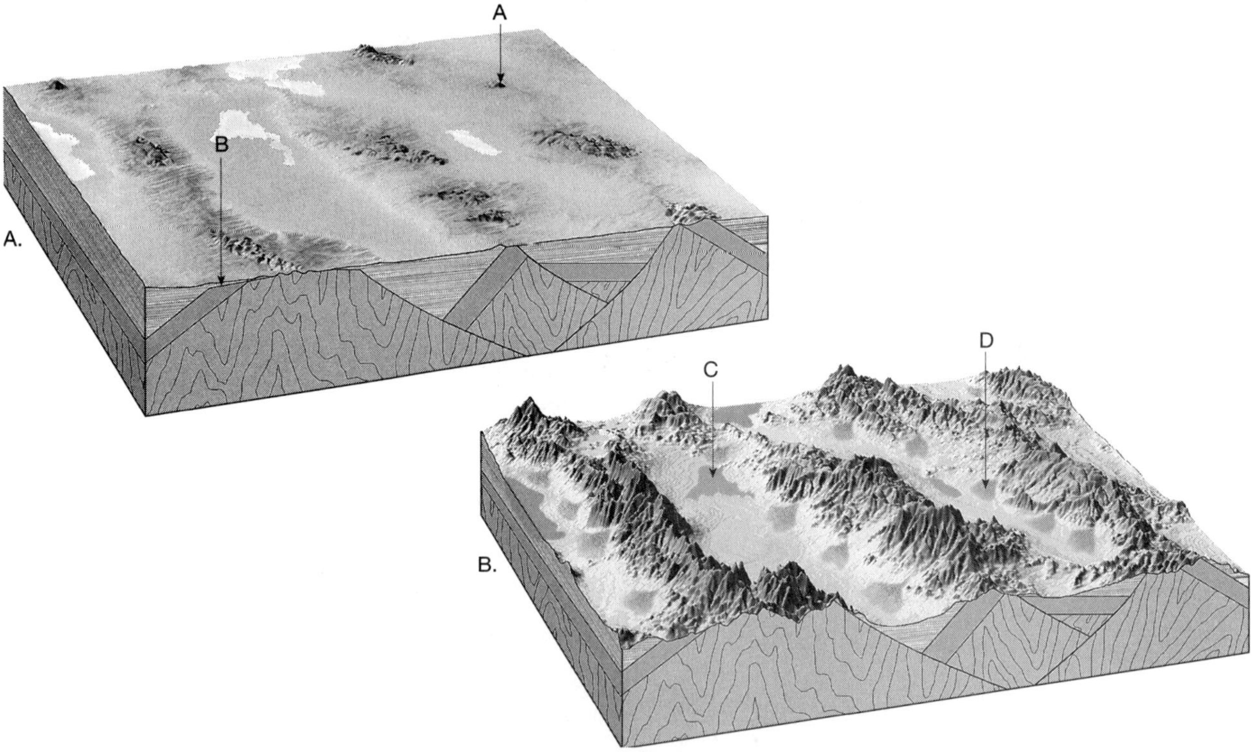

Figure 4.2

8. Distinguish between the following types of moraines:

   a) End moraine:

   b) Lateral moraine:

   c) Ground moraine:

9. The area represented in Figure 4.3 is being subjected to glaciation by an ice sheet. Select the appropriate letter in the figure that identifies each of the following features.

   a) Drumlin: ____ c) Esker: ____

   b) Outwash plain: ____ d) End moraine: ____

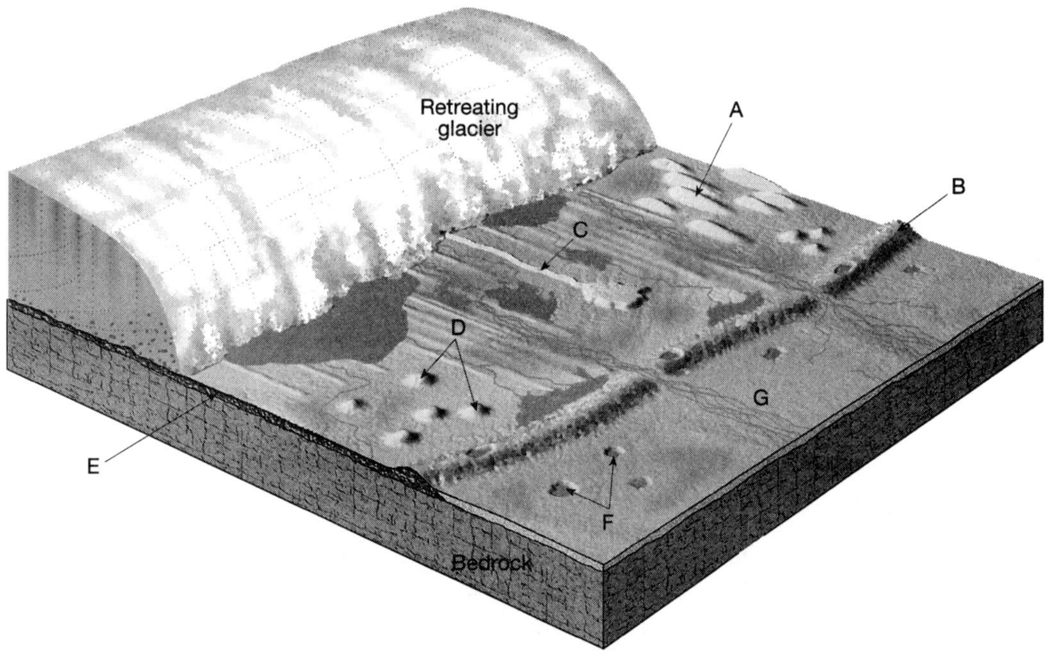

Figure 4.3

10. Briefly describe the process responsible for the migration of sand dunes.

11. What were some indirect effects of Ice Age glaciers?

12. What features would you look for in a mountainous region to determine if the area had been glaciated by alpine glaciers?

13. What evidence suggests that there were several glacial advances during the Ice Age?

14. In what two ways does the transport of sediment by wind differ from that by running water?

Practice Test

Multiple choice. Choose the best answer for the following multiple choice questions.

1. A broad, ramplike surface of stratified drift built adjacent to the downstream edge of most end moraines is called a(n) ____________.
   a) moraine c) outwash plain e) kettle
   b) valley train d) fiord

2. Icebergs are produced when large pieces of ice break off from the front of a glacier during a process termed ____________.
   a) ablation c) calving e) plucking
   b) deflation d) abrasion

3. The dry regions in the lower latitudes coincide with zones of air pressure called the ____________.
   a) equatorial lows c) equatorial doldrums e) subtropical highs
   b) equatorial highs d) subtropical lows

4. Most of the major glacial episodes during the Ice Age occurred during a division of the geologic time scale called the ____________ epoch.
   a) Pliocene c) Paleocene e) Miocene
   b) Eocene d) Pleistocene

5. The two major ways that glaciers erode land are abrasion and ____________.
   a) plucking c) deflation e) scouring
   b) slipping d) gouging

6. Which one of the following is NOT an effect that Pleistocene glaciers had upon the landscape?
   a) mass extinctions
   b) crustal depression and rebounding
   c) animal and plant migration
   d) adjustments of stream courses
   e) worldwide change in sea level

7. A thick ice mass that forms over the land from the accumulation, compaction, and recrystallization of snow is a ___________.
   a) wadi c) drumlin e) loess
   b) fiord d) glacier

8. The ___________ drainage of arid regions is characterized by intermittent streams that do not flow out of the desert to the ocean.
   a) radial c) exterior e) well-developed
   b) interior d) excellent

9. The all-embracing term for sediments of glacial origin is ___________.
   a) outwash c) loess e) erratic
   b) drift d) silt

10. The end moraine that marks the farthest advance of a glacier is called the ___________ moraine.
    a) lateral c) medial e) recessional
    b) terminal d) ground

11. Cracks that form in the zone of fracture of a glacier are called ___________.
    a) fractures c) faults e) kettles
    b) crevasses d) gouges

12. Desert streams that carry water only in response to specific episodes of rainfall are said to be___________.
    a) episodic c) occasional e) youthful
    b) flowing d) ephemeral

13. Sinuous ridges composed of sand and gravel deposited by streams flowing in tunnels beneath glacial ice are ___________.
    a) eskers c) drumlins e) moraines
    b) cirques d) horns

14. Materials that have been deposited directly by a glacier are called ___________.
    a) outwash c) till e) stratified drift
    b) sediment d) loess

15. Valleys that are left standing high above the main trough of a receding valley glacier are termed ___________ valleys.
    a) ephemeral c) glacial e) hanging
    b) wadis d) fiord

16. The thickest and most extensive loess deposits occur in western and northern ___________.
    a) Canada c) Australia e) Egypt
    b) China d) Libya

17. Glacial troughs that have become deep, steep-sided inlets of the sea are called ___________.
    a) fiords c) cirques e) kames
    b) hanging valleys d) washes

18. Moraines that form when two valley glaciers coalesce to form a single ice stream are termed ____________ moraines.
    a) lateral c) terminal e) recessional
    b) medial d) ground

19. Boulders found in glacial till or lying free on the surface are called glacial ____________.
    a) remnants c) drumlins e) erratics
    b) striations d) moraines

20. Which one of the following is NOT a feature associated with valley glaciers?
    a) arête c) horn e) cirque
    b) glacial trough d) arroyo

21. Dry, flat lake beds located in the center of basins in arid areas are called ____________.
    a) playas c) kames e) alluvial fans
    b) arroyos d) deltas

22. The most noticeable result of deflation in some places are shallow depressions called ____________.
    a) kettles c) dunes e) sinkholes
    b) blowouts d) drumlins

23. Evidence indicates that, in addition to the Pleistocene epoch, there were at least ____________ earlier periods of glacial activity.
    a) two c) four e) six
    b) three d) five

24. The leeward slope of a dune is called the ____________ face.
    a) cross c) wind e) gradual
    b) slope d) slip

25. Which one of the following states is include in the Basin and Range region?
    a) Nevada c) Texas e) Florida
    b) Missouri d) New York

True/false. For the following true/false questions, if a statement is not completely true, mark it false. For each false statement, change the ***italicized*** *word to correct the statement.*

1. ___ During the Ice Age, ice sheets and alpine glaciers were far *more* extensive than they are today.

2. ___ Snow accumulation and glacial ice formation occur in an area known as the zone of *wastage.*

3. ___ *Middle* latitude dry regions exist principally in the deep interiors of large landmasses.

4. ___ *Medial* moraines form along the sides of valley glaciers.

5. ___ The two major types of glaciers are *valley* glaciers and ice sheets.

6. ___ Under pressure equivalent to more than the weight of *ten* meters of ice, ice will behave as a plastic and flow.

7. ___ The Ice Age began between two and three *million* years ago.

8. ___ The position of the front of a glacier depends on the balance between *accumulation* and wastage.

9. ___ The combined areas of present-day continental ice sheets represents almost *thirty* percent of Earth's land area.

10. ___ Dry lands are concentrated in the subtropics and *high* latitudes.

11. ___ The rates that glaciers advance vary considerably from one glacier to another but can be as great as several *kilometers* per day.

12. ___ Desert floods arrive suddenly and subside *quickly*.

13. ___ Glaciated valleys typically exhibit a *V-shaped* appearance.

14. ___ Streamlined asymmetrical hills composed of till that occur in clusters are called *eskers*.

15. ___ The movement of glacial ice is generally referred to as *flow*.

16. ___ *Glaciers* are capable of carrying huge blocks of material that no other erosional agent could budge.

17. ___ In *humid* regions, moisture binds particles together and vegetation anchors the soil so that wind erosion is negligible.

18. ___ The Matterhorn in the Swiss Alps is the most famous example of a *cirque*.

19. ___ Sediments laid down by glacial meltwater are called stratified *drift*.

20. ___ *Moraines* are bowl-shaped erosional features found at the heads of glacial valleys.

21. ___ In the United States, *loess* deposits are an indirect product of glaciation.

22. ___ Continued sand accumulation, coupled with periodic slides down the *slip* face, causes the slow migration of a sand dune in the direction of air movement.

Written questions

1. List and briefly describe at least four glacial depositional features.

2. List and briefly describe at least three erosional features you might expect to find in an area where valley glaciers exist, or have recently existed.

3. List three indirect effects of Ice Age glaciers.

4. Describe how sand dunes migrate.

5

PLATE TECTONICS:
A Unifying Theory

Plate Tectonics: A Unifying Theory opens by examining the lines of evidence that Alfred Wegener used in the early 1900s to support his continental drift hypothesis. This evidence included matching the outlines of the shorelines of continents that are now separated by vast ocean basins fossils, rock types and structural similarities between continents, and paleoclimates. Also presented are the main objections to Wegener's ideas.

Following a brief overview, the theory of plate tectonics is examined in detail. The movement of lithospheric plates and different types of plate boundaries are examined extensively. Paleomagnetism (polar wandering and magnetic reversals), the distribution of earthquakes, ages and distribution of ocean basin sediments, and hot spots are used to provide additional support for plate tectonics. The chapter closes with comments about the driving mechanism of plate tectonics.

Learning Objectives

After reading, studying, and discussing this chapter, you should be able to:

- List the evidence that was used to support the continental drift hypothesis.
- Describe the theory of plate tectonics.
- Explain the differences between the continental drift hypothesis and the theory of plate tectonics.
- List and describe the evidence used to support the plate tectonics theory.
- Explain the difference between divergent, convergent, and transform plate boundaries.
- Describe the models that have been proposed to explain the driving mechanism for plate motion.

Chapter Review

- In the early 1900s *Alfred Wegener* set forth his *continental drift* hypothesis. One of its major tenets was that a supercontinent called *Pangaea* began breaking apart into smaller continents about 200 million years ago. The smaller continental fragments then "drifted" to their present positions. To support the claim that the now-separate continents were once joined, Wegener and others used the *fit of South America and Africa, the distribution of ancient climates, fossil evidence*, and *rock structures.*

- One of the main objections to the continental drift hypothesis was its inability to provide an acceptable mechanism for the movement of continents.

- The theory of *plate tectonics*, a far more encompassing theory than continental drift, holds that Earth's rigid outer shell, called the *lithosphere*, consists of about seven large and numerous smaller segments called *plates* that are in motion relative to each other. Most of Earth's *seismic activity, volcanism*, and *mountain building* occur along the dynamic margins of these plates.

• A major departure of the plate tectonics theory from the continental drift hypothesis is that large plates contain both continental and ocean crust and the entire plate moves. By contrast, in continental drift, Wegener proposed that the sturdier continents "drifted" by breaking through the oceanic crust, much like ice breakers cut through ice.

• The three distinct types of plate boundaries are 1) *divergent boundaries*—where plates move apart, 2) *convergent boundaries*—where plates move together as in *oceanic-continental convergence, oceanic-oceanic convergence,* or *continental-continental convergence,* and 3) *transform boundaries*—where plates slide past each other.

• The theory of plate tectonics is supported by 1) *paleomagnetism,* the direction and intensity of Earth's magnetism in the geologic past; 2) the global distribution of *earthquakes* and their close association with plate boundaries; 3) the ages of *sediments* from the floors of the deep-ocean basins; and 4) the existence of island groups that formed over *hot spots* and provide a frame of reference for tracing the direction of plate motion.

• Several models for the driving mechanism of plates have been proposed. One model involves large *convection cells* within the mantle carrying the overlying plates. Another proposes that dense *oceanic material descends and pulls* the lithosphere along. No single driving mechanism can account for all of the major facets of plate motion.

Chapter Outline

- I. Continental drift
  - A. Alfred Wegener
    - 1. First proposed hypothesis, 1915
    - 2. Published *The Origin of Continents and Oceans*
  - B. Supercontinent called Pangaea began breaking apart about 200 million years ago
  - C. Continents "drifted" to present positions
  - D. Continents "broke" through ocean crust
  - E. Evidence used by Wegener and others
    - 1. Fit of South America and Africa
    - 2. Fossils
    - 3. Rock structures
    - 4. Ancient climates
  - F. Objections due to Wegner's: inability to provide a mechanism
- II. Plate tectonics
  - A. More encompassing than continental drift
  - B. Associated with Earth's rigid outer shell
    - 1. Called the lithosphere
    - 2. Consists of several (plates)
      - a. Plates are moving slowly
      - b. Largest plate is the Pacific plate
      - c. Plates are mostly beneath the ocean
  - C. Asthenosphere
    - 1. Exists beneath the lithosphere
    - 2. Hotter and weaker than lithosphere
    - 3. Allows for motion of lithosphere
  - D. Plate boundaries
    - 1. Associated with plate boundaries
      - a. Seismic activity
      - b. Volcanism
      - c. Mountain building
    - 2. Types of plate boundaries
      - a. Divergent (spreading) boundary
        - 1. Most exist along oceanic ridge crests
        - 2. Sea-floor spreading along the boundary forms fractures on ridge crests fractures fill with molten material
        - 3. When the boundary occurs on a continent rifts or rift valleys form
      - b. Convergent boundary
        - 1. Lithosphere subducted into mantle
        - 2. Types of convergent boundaries
          - a. Oceanic-continental boundary
            - 1. Forms a subduction zone
              - a. Deep-ocean trench
              - b. Continental volcanic arcs

1. e.g., Andes
2. e.g., Cascade Range
3. e.g., Sierra Nevada system

b. Oceanic-oceanic boundary
1. Often forms volcanoes on the ocean floor
2. Volcanoes can emerge as a volcanic island arc
   a. e.g., Aleutian islands
   b. e.g., Alaskan Peninsula
   c. e.g., Philippines
   d. e.g., Japan

c. Continental-continental boundary often produces mountains
1. e.g., Himalayas
2. Other possibilities
   a. Alps
   b. Appalachians
   c. Urals

c. Transform boundary
1. Plates slide past one another
   a. No new crust created
   b. No crust destroyed
2. Transform faults
   a. Most in oceanic crust
   b. Join two segments of mid-ocean ridge
   c. Parallel direction of plate movement

E. Evidence that supports plate tectonics
1. Paleomagnetism
   a. Probably most persuasive evidence
   b. Ancient magnetism preserved in rocks
   c. Paleomagnetic records show
      1. Polar wandering—evidence that continents moved
      2. Earth's magnetic field reversals
         a. Recorded in sea-floor rocks
         b. Confirms sea-floor spreading
2. Earthquake patterns
   a. Associated with plate boundaries
   b. Deep-focus earthquakes along trenches provide a method for tracking plate's descent
3. Ocean drilling
   a. Deep Sea Drilling Project (*Glomar Challenger*)
   b. Age of oldest sediments
      1. Youngest near ridges
      2. Older at a distance from ridge
   c. Ocean basins are geologically young
4. Hot spots
   a. Rising plumes of mantle material
   b. Volcanoes can form over them
      1. e.g., Hawaiian Island chain
      2. Chains mark plate movement

F. Driving mechanism of plate tectonics
1. No one model explains all plate motions
2. Earth's heat is the driving force
3. Several models have been proposed
   a. Convection currents in mantle
   b. Slab-pull and slab-push
      1. Descending oceanic crust pulls
      2. Elevated ridge system pushes
   c. Hot plumes
      1. Extend from mantle-core boundary
      2. Spread laterally under lithosphere

Vocabulary Review

Choosing from the list of key terms, furnish the most appropriate response for the following statements.

1. In his 1915 book, *The Origins of Continents and Oceans,* Wegener set forth his radical hypothesis of ___________.

2. Each of the rigid slabs of Earth's lithosphere is referred to as a(n) ___________.

3. ___________ is the mechanism responsible for producing the new sea floor between two diverging plates.

Key Terms

Page numbers shown in ( ) refer to the textbook page where the term first appears.

continental drift (p. 118)
continental volcanic arc (p. 130)
convergent boundary (p. 123)
deep-ocean trench (p. 128)
divergent boundary (p. 123)
hot spot (p. 138)
normal polarity (p. 135)
paleomagnetism (p. 134)
Pangaea (p. 118)
plate (p. 122)
plate tectonics (p. 122)
polar wandering (p. 134)
reverse polarity (p. 135)
rift (rift valley) (p. 126)
sea floor spreading (p. 123)
subduction zone (p. 127)
transform boundary (p. 123)
volcanic island arc (p. 130)

4. The type of plate boundary where plates move together, causing one of the slabs to be consumed into the mantle as it descends beneath an overriding plate, is referred to as a(n) ____________.

5. Remnant magnetism in rock bodies is referred to as ____________.

6. The theory of ____________ holds that Earth's rigid outer shell consists of about twenty rigid slabs.

7. When rocks exhibit the same magnetism as the present magnetic field, they are said to possess____________.

8. The type of plate boundary where plates move apart, resulting in upwelling of material from the mantle to create new sea floor, is referred to as a(n) ____________.

9. A rising plume of mantle material often causes a(n) ____________, such as the one responsible for the intraplate volcanism that produced the Hawaiian Islands.

10. Wegener hypothesized that about 200 million years ago a supercontinent that he called____________ began breaking into smaller continents, which then drifted to their present positions.

11. The region where an oceanic plate descends into the asthenosphere because of convergence is called a(n) ____________.

12. Material displaced downward along spreading centers on continents often creates a down-faulted valley called a(n) ____________.

13. A(n) ____________ occurs where plates grind past each other without creating or destroying lithosphere.

14. ____________ is the idea that Earth's magnetic poles have migrated greatly through time.

15. Rocks that exhibit magnetism that is opposite of the present magnetic field are said to possess ____________.

16. A chain of small volcanic islands that forms when two oceanic slabs converge, one descending beneath the other, is called a(n) ____________.

17. As an oceanic plate slides beneath an overriding plate, a deep, linear feature, called a(n) ____________, often forms on the ocean floor adjacent to the zone of subduction.

Comprehensive Review

1. List four lines of evidence that were used to support the continental drift hypothesis.
   1)

   2)

   3)

   4)

2. What was one of the main objections to Wegener's continental drift hypothesis?

3. Using Figure 5.1, select the letter of the diagram that portrays each of the following types of plate boundaries.
   a) Transform boundary: ____
   b) Divergent boundary: ____
   c) Convergent boundary: ____

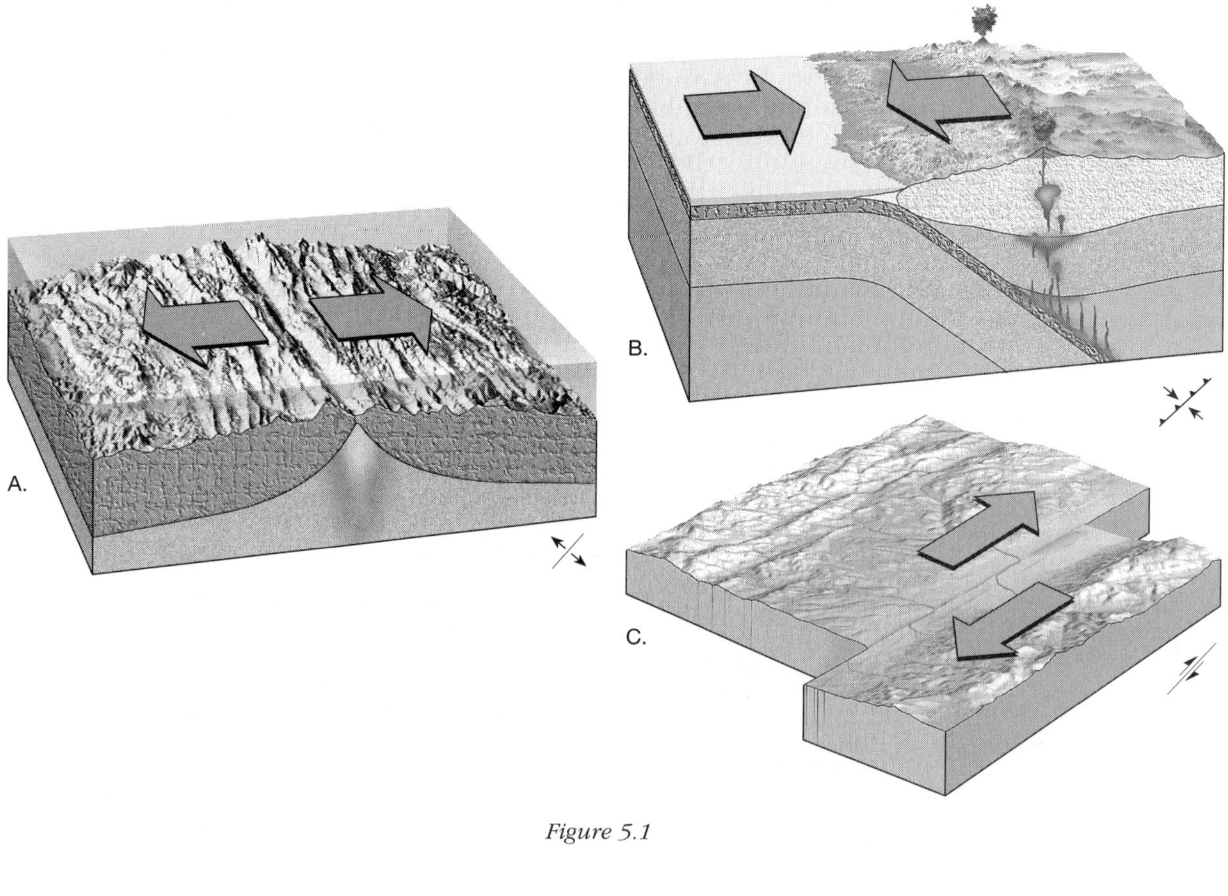

Figure 5.1

4. Briefly explain the theory of plate tectonics.

5. What are the three types of convergent plate boundaries?

   1)

   2)

   3)

6. Briefly explain how each of the following has been used to support the theory of plate tectonics:
   a) Paleomagnetism:

   b) Earthquake patterns:

   c) Ocean drilling:

7. Using Figure 5.2, select the appropriate letter that identifies each of the following features.
   a) Subducting oceanic lithosphere: ____
   b) Trench: ____
   c) Continental volcanic arc: ____
   d) Continental lithosphere: ____

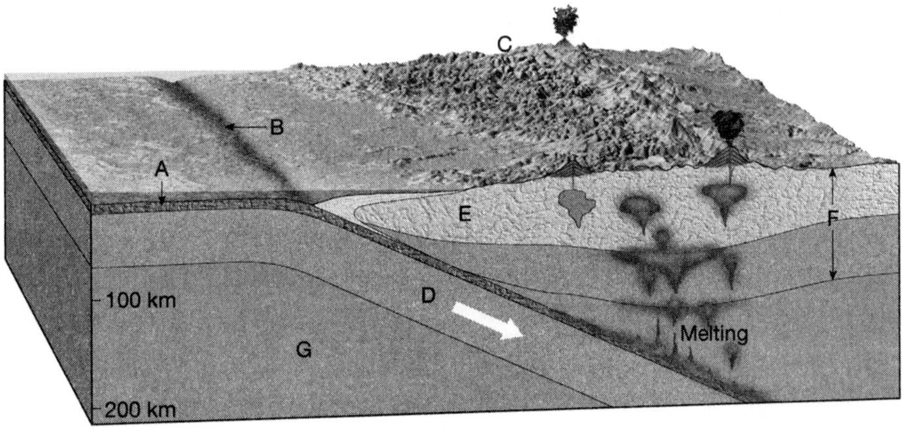

Figure 5.2

8. List and briefly describe the three hypotheses that have been proposed for the driving mechanism of plate motion.

   1)

   2)

   3)

9. Figure 5.3 illustrates the ancient supercontinent of Pangaea as it is thought to have existed about 300 million years ago. Using the figure, select the appropriate letter that identifies each of the following current-day landmasses.

   a) Antarctica: ____

   b) Eurasia: ____

   c) South America: ____

   d) India: ____

   e) North America: ____

   f) Africa: ____

   g) Australia: ____

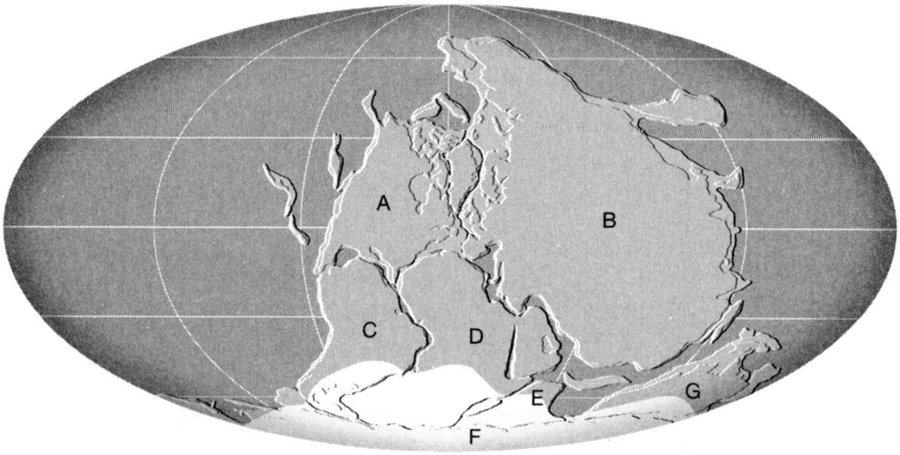

Figure 5.3

Practice Test

Multiple choice. Choose the best answer for the following multiple choice questions.

1. The general term that refers to the deformation of Earth's crust and results in the formation of structural features such as mountains is ___________.
   a) erosion c) mass wasting e) subduction
   b) tectonics d) volcanism

2. During the last four million years, Earth's magnetic field has reversed ___________.
   a) twice b) several times
   c) hundreds of times d) thousands of times

3. Most of Earth's seismic activity, volcanism, and mountain building occur along ___________.
   a) lines of magnetism c) parallels of latitude e) hot spots
   b) plate boundaries d) random trends

4. The apparent movement of Earth's magnetic poles over time is referred to as ___________.
   a) magnetic drift c) subduction e) polar wandering
   b) perturbation d) polar spreading

5. Alfred Wegener is best known for his hypothesis of ___________.
   a) continental drift c) atoll formation e) sea floor spreading
   b) natural selection d) subduction

6. Complex mountain systems such as the Alps, Appalachians, and Himalayas are the result of ___________.
   a) oceanic-oceanic convergence
   b) hot spots
   c) oceanic-continental convergence
   d) continental-continental convergence
   e) island arcs

7. The type of plate boundary where plates move together, causing one of the slabs of lithosphere to be consumed into the mantle as it descends beneath an overriding plate, is called a ___________ boundary.
   a) divergent c) transform e) convergent
   b) transitional d) gradational

8. The typical rate of spreading along ridges in the Atlantic Ocean is estimated to be from 2 to _________ centimeters.
   a) 5 c) 15 e) 25
   b) 10 d) 20

9. The name given by Wegener to the supercontinent he believed existed prior to the current continents was ___________.
   a) Pangaea c) Euroamerica e) Panamerica
   b) Atlantis d) Pantheon

10. Magnetic minerals lose their magnetism when heated above a certain temperature, called the ____________.
    a) magnetopoint c) melting point e) Bechtol point
    b) Curie point d) break point

11. The best approximation of the true outer boundary of the continents is the seaward edge of the ____________.
    a) deep-ocean trench c) mid-ocean ridge e) continental shelf
    b) abyssal plain d) present-day shorelines

12. The Aleutian, Mariana, and Tonga islands are examples of ____________.
    a) island arcs c) transform boundaries e) mid-ocean ridges
    b) abyssal plains d) hot spots

13. Beneath Earth's lithosphere, the hotter, weaker zone known as the ____________ allows for motion of Earth's rigid outer shell.
    a) crust c) outer core e) oceanic crust
    b) asthenosphere d) Moho

14. Mountains comparable in age and structure to the mountain belt that contains the Appalachians are found in ____________.
    a) Africa c) the British Isles e) Asia
    b) Australia d) South America

15. Which one of the following was NOT used in support of Wegener's continental drift hypothesis?
    a) fossil evidence c) fit of South America and Africa e) rock structures
    b) paleomagnetism d) ancient climates

16. The Red Sea is believed to be the site of a recently formed ____________.
    a) divergent boundary c) convergent boundary e) gradational boundary
    b) ocean trench d) hot spot

17. The type of plate boundary where plates move apart, resulting in upwelling of material from the mantle to create new sea floor is a ____________ boundary.
    a) divergent c) transform e) convergent
    b) transitional d) gradational

18. One of the main objections to Wegener's hypothesis was his inability to provide an acceptable ____________ for continental drift.
    a) time c) mechanism e) place
    b) rate d) direction

19. The spreading rate for the North Atlantic Ridge is ____________ than the rates for the East Pacific Rise.
    a) greater b) less

20. Earth's rigid outer shell is called the ____________.
    a) asthenosphere c) outer core e) lithosphere
    b) mantle d) continental mass

21. Most deep-focus earthquakes occur in association with ___________.
    a) hot spots   c) oceanic ridge systems   e) transform boundaries
    b) ocean trenches   d) abyssal plains

22. The theory of plate tectonics holds that Earth's rigid outer shell consists of about ___________ large and numerous smaller rigid slabs.
    a) five   c) fifteen   e) twenty–five
    b) seven   d) twenty

23. The age of the deepest sediment in an ocean basin ___________ with increasing distance-from the oceanic ridge.
    a) increases   b) decreases   c) remains the same   d) varies

24. During oceanic-continental convergence, as the oceanic plate slides beneath the overriding plate, a(n) ___________ is often produced adjacent to the zone of subduction.
    a) deep-ocean terrace   c) transform fault   e) divergent boundary
    b) deep-ocean ridge   d) deep-ocean trench

25. The chain of volcanic structures that extends from the Hawaiian Islands to Midway Island and then continuing northward toward the Aleutian trench have formed over a(n) ___________ as the Pacific plate moved.
    a) subduction zone   c) island arc   e) divergent boundary
    b) hot spot   d) convergent boundary

26. Which one of the following is NOT a hypothesis that has been proposed for the mechanism of plate motion?
    a) hot plumes hypothesis   c) convection current hypothesis
    b) slab-push and slab-pull hypothesis   d) mantle density hypothesis

True/false. For the following true/false questions, if a statement is not completely true, mark it false. For each false statement, change the ***italicized*** *word to correct the statement.*

1. ___ The *lithosphere* consists of both crustal rocks and a portion of the upper mantle.
2 .___ *Transform* faults join two segments of a mid-ocean ridge.
3. ___ The rate of plate movement is measured in *kilometers* per year.
4. ___ The Red Sea is believed to the site of a recently formed *convergent* plate boundary.
5. ___ In North America, the *Cascade Range* and Sierra Nevada system are volcanic arcs associated with the subduction of oceanic lithosphere.
6. ___ Older portions of the sea floor are carried into Earth's *core* in regions where trenches occur in the deep ocean floor.
7. ___ The island of Hawaii is *older* than Midway Island.
8. ___ The "Ring of Fire" is an area of earthquake and volcanic activity that encircles the *Pacific* ocean basin.
9. ___ *Transform* faults are roughly parallel to the direction of plate movement.
10. ___ To explain continental drift, Wegener proposed that the *continents* broke through the oceanic crust, much like ice breakers cut through ice.

11. ___ There is a close association between deep-focus earthquakes and ocean *ridges.*

12. ___ Beneath Earth's lithosphere is the hotter and weaker zone known as the *asthenosphere.*

13. ___ Sea-floor spreading is the mechanism that has produced the floor of the *Atlantic* Ocean during the past 165 million years.

14. ___ The unequal distribution of heat inside Earth generates some type of thermal convection in the *crust* which ultimately drives plate motion.

15. ___ The supercontinent of *Pangaea* began breaking apart about 200 million years ago.

16. ___ When rocks exhibit the same magnetism as the present magnetic field they are said to possess *reverse* polarity.

17. ___ Deep-ocean trenches are located adjacent to *subduction* zones.

18. ___ The *oldest* oceanic crust is located at the oceanic ridge crests.

19. ___ The largest single rigid slab of Earth's outer shell is the *Pacific* plate.

20. ___ The oldest sediments found in the ocean basins are *less* than 160 million years old.

21. ___ Along a *transform* plate boundary, plates grind past each other without creating or destroying lithosphere.

22. ___ The Aleutian, Mariana, and Tonga islands are volcanic island arcs associated with *oceanic-oceanic* plate convergence.

23. ___ Lithospheric plates are *thickest* in the ocean basins.

24. ___ The idea that Earth's *magnetic* poles had migrated through time is known as polar wandering.

25. ___ Using the dates of the most recent *magnetic* reversals, the rate at which spreading occurs at the various ridges can be determined.

Written questions

1. What relation exists between the ages of the Hawaiian Islands, hot spots, and plate tectonics?

2. List and briefly describe the three major types of plate boundaries.

3. List the lines of evidence used to support the theory of plate tectonics.

RESTLESS EARTH:

Earthquakes, Geologic Structures, and Mountain Building

6

Restless Earth: Earthquakes, Geologic Structures, and *Mountain Building* begins with a brief description of the effects of the major earthquakes that have taken place in California within the last decade. Following an explanation of how earthquakes occur, the types of seismic waves, their propagation, and how they appear on a typical seismic trace are presented. This is followed by a discussion of earthquake epicenters: how they are located and their worldwide distribution. Earthquake magnitude is also explained. The destruction caused by seismic vibrations and their associated perils introduces a discussion of earthquake prediction. The presentation of earthquakes closes with an explanation of how earthquakes are used to investigate Earth's interior structure as well as a general description of its composition.

Geologic structures and mountain building opens with a brief explanation of the various types of folds (anticlines, synclines, domes, and basins) and faults (both dip-slip and strike-slip). The chapter concludes with an extensive presentation of mountain building and its association with plate boundaries.

Learning Objectives

After reading, studying, and discussing this chapter, you should be able to:

- Describe the cause of earthquakes.
- List the types of seismic waves and describe their propagation.
- Describe how an earthquake epicenter is located.
- Describe the worldwide distribution of earthquake epicenters.
- Explain how the magnitude of an earthquake is determined.
- List the other destructive forces that can be triggered by an earthquake.
- Discuss the status of earthquake prediction.
- Describe Earth's interior structure and composition.
- List the major types of folds and faults and describe how they form.
- Describe the relation between mountain building and plate tectonics.

Chapter Review

- *Earthquakes* are vibrations of Earth produced by the rapid release of energy from rocks that rupture because they have been subjected to stresses beyond their limit. This energy, which takes the form of waves, radiates in all directions from the earthquake's source, called the *focus.* The movements that produce most earthquakes occur along large fractures, called *faults,* that are associated with plate boundaries.

• Two main groups of *seismic waves* are generated during an earthquake: 1) *surface waves*, which travel along the outer layer of Earth; and 2) *body waves*, which travel through Earth's interior. Bodywaves are further divided into *primary*, or *P*, *waves*, which push (compress) and pull (dilate) rocks in the direction the wave is traveling, and *secondary*, or *S*, *waves*, which "shake" the particles in rock at right angles to their direction of travel. P waves can travel through solids, liquids, and gases. Fluids (gases and liquids) will not transmit S waves. In any solid material, P waves travel about 1.7 timesfaster than S waves.

• The location on Earth's surface directly above the focus of an earthquake is the *epicenter*. An epicenter is determined using the difference in velocities of P and S waves.

• *There is a close correlation between earthquake epicenters and plate boundaries.* The principal earthquake epicenter zones are along the outer margin of the Pacific Ocean, known as the circum-Pacific belt, and through the world's oceans along the *oceanic ridge system.*

• Using the *Richter scale*, the *magnitude* (a measure of the total amount of energy released) of an earthquake is determined by measuring the *amplitude* (maximum displacement) of the largest seismic wave recorded, with adjustments of the amplitude made for the weakening of seismic waves as they move from the focus, as well as for the sensitivity of the recording instrument. A logarithmic scale is used to express magnitude, in which a tenfold increase in recorded wave amplitude corresponds to an increase of one on the magnitude scale.

• The most obvious factors that determine the amount of destruction accompanying an earthquake are the *magnitude* of the earthquake and the *proximity* of the quake to a populated area. *Structural damage* attributable to earthquake vibrations depends on several factors, including 1) *intensity*, 2) *duration* of the vibrations, 3) *nature of the material* upon which the structure rests, and 4) the *design* of the structure. Secondary effects of earthquakes include *tsunamis*, *fires*, *landslides*, and *ground subsidence*.

• As indicated by the behavior of P and S waves as they travel through Earth, the four major zones of Earth's interior are 1) *crust* (the very thin outer layer), 2) *mantle* (a rocky layer located belowthe crust with a thickness of 2885 kilometers), 3) *outer core* (a layer about 2270 kilometers thick,which exhibits the characteristics of a mobile liquid), and 4) *inner core* (a solid metallic sphere with a radius of about 1216 kilometers).

• The *continental crust* is primarily made of *granitic* rocks, while the *oceanic crust* is of *basaltic* composition. *Ultramafic* rocks such as *peridotite* are thought to make up the *mantle*. The core iscomposed mainly of *iron and nickel*.

• The most basic geologic structures associated with rock deformation are *folds* (flat-lying sedimentary and volcanic rocks bent into a series of wavelike undulations) and *faults* (fractures in the crust along which appreciable displacement has occurred). The two most common typesof folds are *anticlines*, formed by the upfolding, or arching, of rock layers, and *synclines*, which are downfolds, or troughs. Faults in which the movement is primarily vertical are called *dip-slip faults*. Dip-slip faults include both *normal* and *reverse faults*. In *strike-slip faults*, horizontal movement causes displacement along the trend, or strike, of the fault.

• *Major mountain systems form along convergent plate boundaries. Andean-type mountain building* along continental margins involves the convergence of an oceanic plate and a plate whose leading edge contains continental crust. At some point in the formation of Andean-type mountains a subduction zone forms. *Continental collisions*, in which both plates may be carrying continental crust, have resulted in the formation of the Himalaya Mountains and Tibetan Highlands. Recent investigations indicate that *accretion*, a third mechanism of *orogenesis* (the processes that collectively result in the formation of mountains), takes place where *smaller crustal fragments collide and merge with continental margins* along some plate boundaries. Many of the mountainous regions rimming the Pacific have formed in this manner.

Chapter Outline

- I. Earthquakes
  - A. General features
    - 1. Vibration of Earth by release of energy
    - 2. Associated with movements along faults
      - a. Explained by plate tectonics theory
      - b. Mechanism first explained by H. Reid
        - 1. Early 1900s
        - 2. Rocks "spring back" in a phenomena called elastic rebound
      - c. Fault creep
    - 3. Often preceded by foreshocks
    - 4. Often followed by aftershocks
  - B. Earthquake waves
    - 1. Study of earthquake waves is called seismology
    - 2. Recording instrument called a seismograph records a seismogram
    - 3. Types of earthquake waves
      - a. Surface waves
        - 1. Complex motion
        - 2. Slowest velocity of all waves
      - b. Body waves
        - 1. Primary (P) waves
          - a. Push-pull (compressional) motion
          - b. Travel through
            - 1. Solids
            - 2. Liquids
            - 3. Gases
          - c. Greatest velocity of all waves
        - 2. Secondary (S) waves
          - a. "Shake" motion
          - b. Travel only through solids
          - c. Slower velocity than P waves
  - C. Locating an earthquake
    - 1. Focus—place within Earth, where waves originate
    - 2. Epicenter
      - a. On surface, directly above focus
      - b. Located using P and S wave recordings
        - 1. Difference in arrival times
          - a. Noted on seismogram, and
          - b. Related to distance
          - c. Three station recordings needed to find
    - 3. Earthquake zones have a close correlation with plate boundaries
      - a. Circum-Pacific belt
      - b. Oceanic ridge system
  - D. Earthquake magnitude
    - 1. Introduced by Charles Richter, 1935
    - 2. Richter scale
      - a. Magnitude
        - 1. Amplitude of largest wave recorded
        - 2. Largest quakes near magnitude 8.6
        - 3. Less than 2.0 usually not felt
        - 4. Each unit of magnitude increase is about a 30-fold energy increase
  - E. Earthquake destruction
    - 1. Factors that determine destruction
      - a. Magnitude of earthquake
      - b. Proximity to population
    - 2. Destruction from
      - a. Ground shaking
      - b. Liquefaction where saturated material turns fluid
      - c. Tsunamis, or seismic sea waves
      - d. Fires
      - e. Landslides and ground subsidence
  - F. Prediction
    - 1. Short-range—no reliable method yet devised
    - 2. Long-range
      - a. Premise is that earthquakes are repetitive
      - b. Region is given a probability of a quake
- II. Earth's interior
  - A. Most knowledge of interior comes from study of P and S waves
    - 1. Travel times of P and S waves through Earth vary
      - a. Travel times depend on properties of materials
      - b. P waves travel fastest
    - 2. S waves travel only through solids
  - B. Earth's interior structure
    - 1. Crust
      - a. Thin outer layer
      - b. Varies in thickness
        - 1. 70 km in some mountainous regions

2. Less than 5 km in oceanic regions
c. Two parts
1. Continental crust—lighter, granitic rocks
2. Oceanic crust—basaltic composition
d. Lithosphere
1. Crust and uppermost mantle (100 km thick)
2. Cool, rigid, solid
e. Mohorovicic discontinuity (Moho)
1. Boundary that separates crust from mantle
2. Mantle
a. Below crust
b. 2885 km thick
c. Composition similar to the rock peridotite
d. Asthenosphere
1. Upper mantle
2. At a depth from 100 km to 350 km
3. Hot, weak rock
4. Easily deformed
5. Up to 10% is molten
6. Key to explaining plate movement
3. Outer core
a. Below mantle
b. 2270 km thick
c. Mobile liquid—S waves do not travel through
d. Mainly iron and nickel composition
e. Related to Earth's magnetic field
4. Inner core
a. 1216 km radius
b. Solid
c. Iron and nickel composition
d. High density

III. Geologic structures
A. Two basic types of geologic structures
1. Folds
2. Faults
B. Folds
1. Rocks bent into a series of waves
2. Types of common folds
a. Anticline—upfolded, or arched, rock layers
b. Syncline—downfolded, or trough, rock layers
3. A fold can be
a. Symmetrical
b. Asymmetrical
c. Overturned
4. Most folds form from compressional stresses that shorten and thicken the crust
5. Other types
a. Dome
1. Circular, or slightly elongated
2. Upwarped displacement of rocks
3. Oldest rocks in core
b. Basin
1. Circular, or slightly elongated
2. Downwarped displacement of rocks
3. Youngest rocks in core
C. Faults
1. Fractures (breaks) in rocks with displacement
2. Types of faults
a. Dip-slip fault
1. Movement along tilt (dip) of fault
2. Parts of a dip-slip fault
a. Hanging wall (rock above fault)
b. Footwall (rock below fault)
3. Types of dip-slip faults
a. Normal fault
1. Tensional force—pulls rock apart
2. Hanging wall moves down
b. Reverse fault
1. Compressional force—squeezes rock
2. Hanging wall moves up
c. Thrust fault
1. Strong compressional forces
2. Hanging wall moves up
3. Low-angle reverse fault
4. More horizontal
b. Strike-slip fault
1. Movement along trend, or strike
2. Transform fault
a. Large strike-slip fault
b. Often associated with plate boundaries

3. Pulling of crust apart (tension) can form
   a. Graben
      1. Down-dropped block
      2. Bounded by normal faults
      3. Can produce an elongated valley (e.g., Great Rift Valley, Africa)
   b. Horst
      1. Flanking a graben
      2. Relatively uplifted block

IV. Mountain building
   A. Refers to processes that uplift mountains
   B. Major mountain systems are related to plate tectonics and most form along convergent plates
   C. Associated with
      1. Convergent boundaries
         a. Oceanic-continental crust convergence
            1. Andean-type mountains
            2. Subduction zone forms
            3. Deformation of continental margin
            4. Volcanic arc forms
            5. Accretionary wedge
            6. Examples
         b. Where continental crusts converge
            1. e.g., India and Eurasian plate collision
               a. Himalaya Mountains
               b. Tibetan Highlands
            2. Other examples
               a. Alps
               b. Appalachians
      2. Continental accretion
         a. Third mechanism of mountain building
         b. Small crustal fragments collide with and accrete to continental margins
         c. Accreted crustal blocks are called terranes
         d. Occurred along Pacific Coast

Vocabulary Review

Choosing from the list of key terms, furnish the most appropriate response for the following statements.

1. The vibration of Earth produced by a rapid release of energy is called a(n) ____________.

2. A large ocean wave that often results from vertical displacement of the ocean floor duringan earthquake is called a(n) ____________.

3. A small earthquake called a(n) ____________ often precedes a major earthquake by days or, in some cases, by as much as several years.

4. The source of an earthquake is called the ____________.

5. The "springing back" of rock after it has been deformed, much like a stretched rubber band does when released, is termed ____________.

6. The type of earthquake wave that travels along the outer layer of Earth is the ____________.

7. A fault in which the dominant displacement is along the trend of the fault is called a(n) ____________.

8. The ____________ of an earthquake is the location on the surface directly above the focus.

9. The zone between about 100 kilometers and 350 kilometers within Earth that consists of hot, weak rock that is easily deformed is called the ____________.

Key Terms

Page numbers shown in ( ) refer to the textbook page where the term first appears.

accretionary wedge (p. 169)
aftershock (p. 149)
anticline (p. 165)
asthenosphere (p. 163)
basin (p. 165)
body wave (p. 150)
crust (p. 161)
dip-slip fault (p. 165)
dome (p. 165)
earthquake (p. 147)
elastic rebound (p. 148)
epicenter (p. 151)
fault (p. 147)
fault creep (p. 148)
focus (p. 147)
fold (p. 165)
foreshock (p. 149)
graben (p. 167)
horst (p. 167)
inner core (p. 161)
liquefaction (p. 156)
lithosphere (p. 163)
magnitude (p. 152)
mantle (p. 161)
Mohorovicic discontinuity (Moho) (p. 161)
mountain building (p. 168)
normal fault (p. 166)
outer core (p. 161)
primary (P) wave (p. 150)
reverse fault (p. 166)
Richter scale (p. 152)
secondary (S) wave (p. 150)
seismic sea wave (tsunami) (p. 156)
seismogram (p. 150)
seismograph (p. 149)
seismology (p. 149)
shadow zone (p. 162)
strike-slip fault (p. 166)
surface wave (p. 150)
syncline (p. 165)
terrane (p. 173)
thrust fault (p. 166)

10. A fracture in rock along which displacement has occurred is termed a(n) ___________.
11. A(n) ___________ is the type of fold most commonly formed by the upfolding, or arching, of rock layers.
12. The type of earthquake wave that "shakes" the particles at right angles to the direction it is traveling is the ___________.
13. A recording, or trace, of an earthquake is called a(n) ___________.
14. The boundary that separates the crust from the mantle is known as the ___________.
15. Using Richter's scale, the ___________ of an earthquake is determined by measuring the amplitude of the largest wave recorded on the seismogram.
16. Earth's ___________ is a solid metallic sphere about 1216 kilometers in radius.
17. The type of earthquake wave that pushes and pulls rock in the direction it is traveling is the ___________.
18. The cool, rigid layer of Earth, which includes the entire crust as well as the uppermost mantle, is the ___________.
19. A downfold, or trough, of rock layers forms a type of fold called a(n) ___________.
20. A dip-slip fault is classified as a(n) ___________ when the hanging wall moves down relative to the footwall.
21. The rocky layer of Earth located beneath the crust and having a thickness of 2885 kilometers is the ___________.

22. An accumulation of sedimentary and metamorphic rocks, including occasional scraps of ocean crust, that forms in association with a subduction zone is called a(n) ____________.

23. A crustal block whose geologic history is distinct from the histories of adjoining crustal blocks is termed a(n) ____________.

24. Gradual displacement known as ____________ occurs along the San Andreas fault with little noticeable seismic activity.

Comprehensive Review

1. List and describe the three basic types of earthquake waves.

   1)

   2)

   3)

2. Using Figure 6.1, select the letter of the diagram that illustrates the characteristic motion of the following types of earthquake waves.

   a) P wave: ____ b) S wave: ____

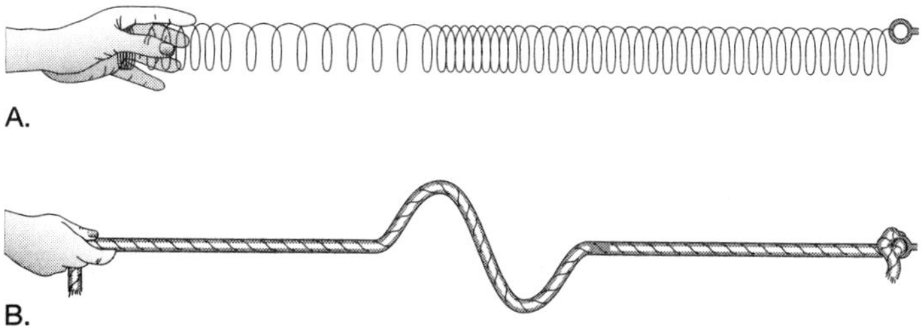

Figure 6.1

3. Figure 6.2 is a typical recording of an earthquake, called a seismogram. Using Figure 6.2, select the letter on the figure that identifies the recording of each of the following types of earthquake waves.

   a) Surface waves: ____

   b) S waves: ____

   c) P waves: ____

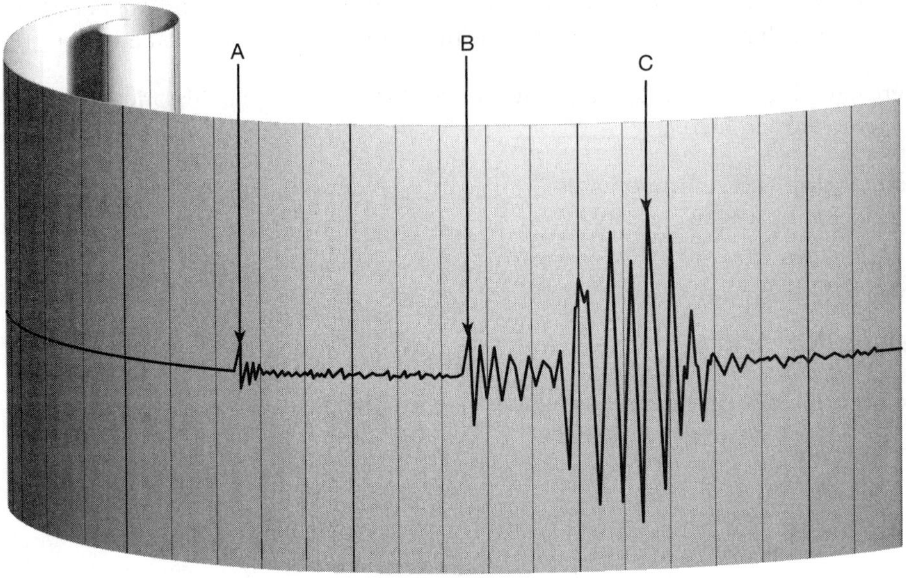

Figure 6.2

4. Describe the differences in velocity and mode of travel between P waves and S waves.

5. Briefly describe how the epicenter of an earthquake is located.

6. What are two factors that determine the degree of destruction that accompanies an earthquake?

   1)

   2)

7. List two major, continuous earthquake belts. With what general Earth feature are most earthquake epicenters closely correlated?

8. Using Figure 6.3, select the letter that identifies each of the following parts of Earth's interior.

a) Mantle: ____

b) Inner core: ____

c) Continental crust: ____

d) Oceanic crust: ____

e) Asthenosphere: ____

f) Outer core: ____

g) Lithosphere: ____

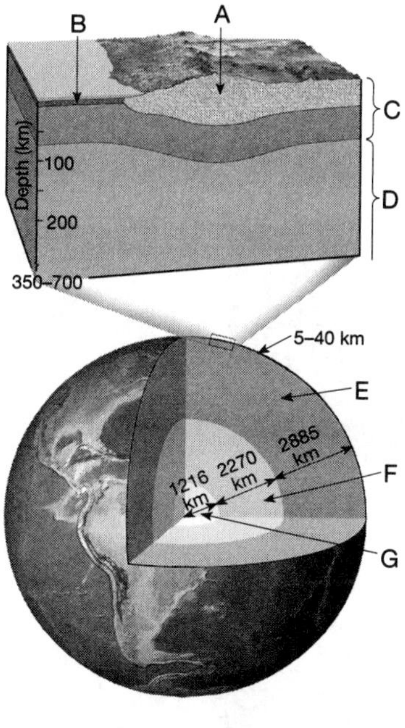

Figure 6.3

9. Briefly describe the composition of the following zones of Earth's interior.

   a) Mantle:

   b) Outer core:

10. Briefly describe each of the following types of faults and select the letter of the diagram in Figure 6.4 that illustrates it.

    a) Reverse fault: ____

    b) Strike-slip fault: ____

    c) Normal fault: ____

    d) Thrust fault: ____

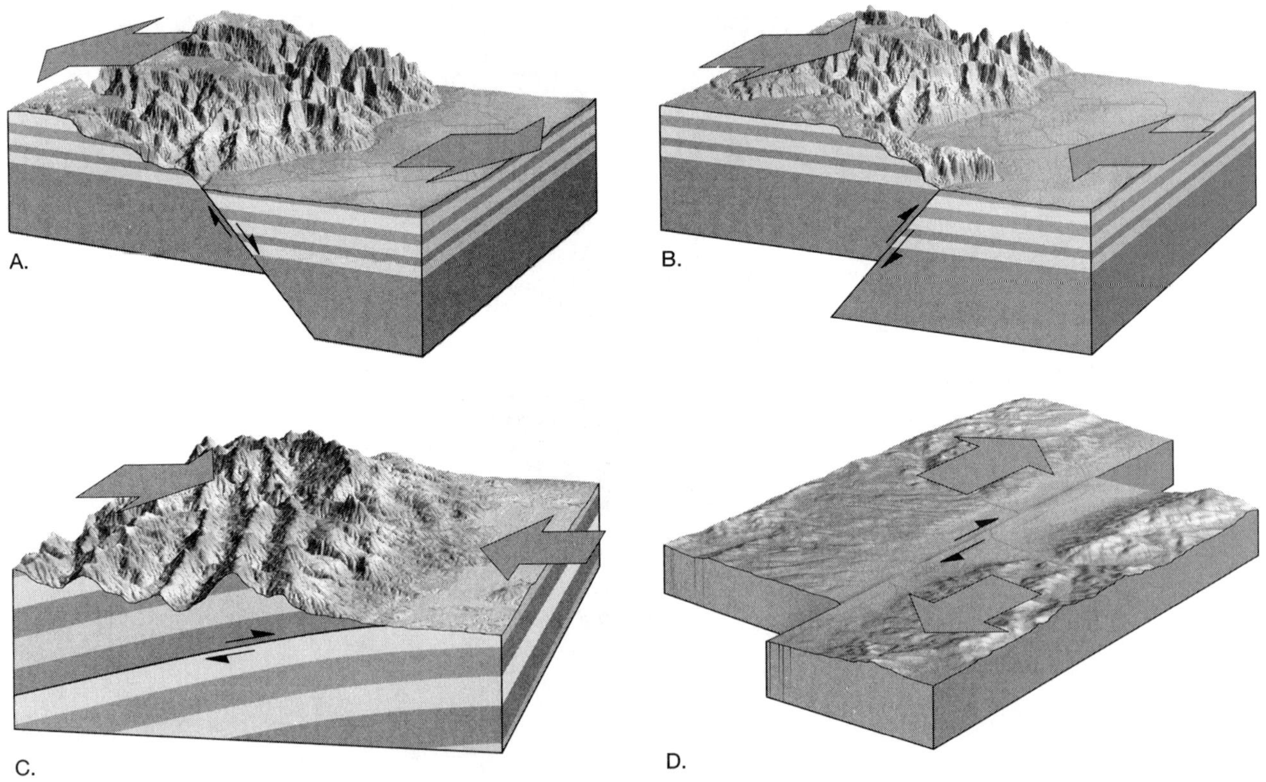

Figure 6.4

11. Using Figure 6.5, select the letter that illustrates a

a) Graben: ____ b) Horst: ____

12. Which type of fault are those illustrated in Figure 6.5?

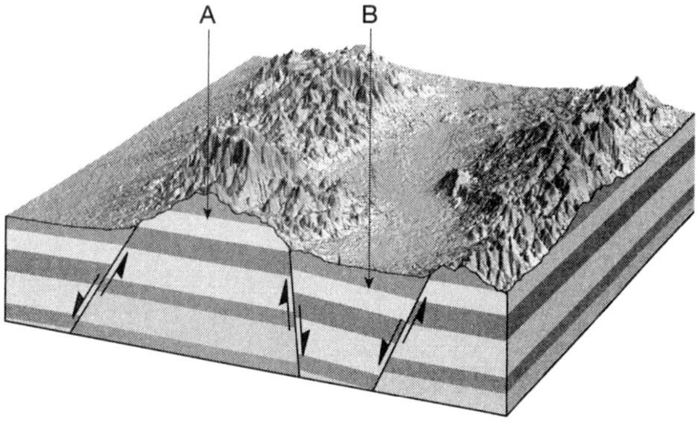

Figure 6.5

13. Briefly describe each of the following types of folds and select the letter of the diagram in Figure 6.6 that illustrates it.

a) Anticline: ____ b) Syncline: ____

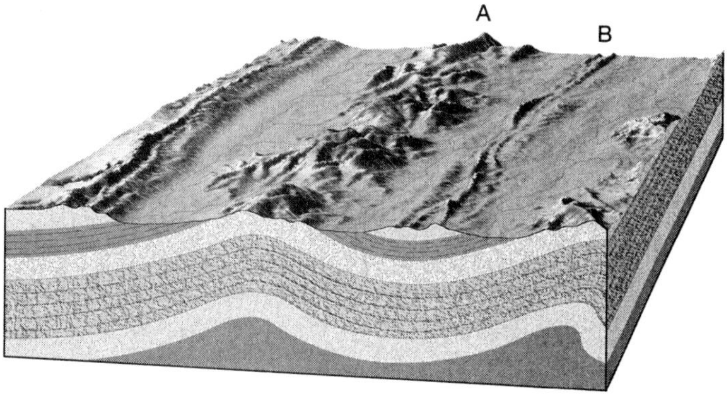

Figure 6.6

14. Briefly explain the events that produce mountains at convergent boundaries where

a) Oceanic and continental crusts converge:

b) Continental crusts converge:

Practice Test

Multiple choice. Choose the best answer for the following multiple choice questions.

1. It is estimated that over ____________ earthquakes that are strong enough to be felt occur worldwide annually.
   a) 500 b) 1000 c) 10000 d) 20000 e) 30000

2. The location on the surface directly above the earthquake focus is called the ____________.
   a) ephemeral b) epicycle c) epicenter d) epinode e) epitaph

3. The cool, rigid layer of Earth that includes the entire crust as well as the uppermost mantle is called the ____________.
   a) asthenosphere b) lower crust c) oceanic crust d) Moho e) lithosphere

4. The two most common types of folds are anticlines and ____________.
   a) domes b) synclines c) basins d) superclines e) batholiths

5. Which earthquake body wave has the greatest velocity?
   a) P wave b) S wave

6. The belt from about 105 to 140 degrees away from an earthquake where no P waves are recorded is known as the ____________.
   a) shadow zone b) absent zone c) Moho zone d) reflective zone e) low velocity zone

7. The study of earthquakes is called ____________.
   a) seismogram b) seismicity c) seismology d) seismography e) seismogony

8. The rock immediately above a fault surface is commonly called the ____________.
   a) anticline b) foot-wall c) syncline d) hanging wall e) dip-slip

9. The difference in ____________ of P and S waves provides a method for determining the epicenter of an earthquake.
   a) magnitudes b) velocities c) sizes d) modes of travel e) foci

10. The source of an earthquake is called the ____________.
    a) fulcrum b) ephemeral c) epicenter d) foreshock e) focus

11. Long-range earthquake forecasts are based on the premise that earthquakes are ____________.
    a) random b) destructive c) fully understood d) always occurring e) repetitive

12. The epicenter of an earthquake is located using the distances from a minimum of ____________ seismic stations.
    a) three c) five e) seven
    b) four d) six

13. Which of the earthquake body waves cannot be transmitted through fluids?
    a) P waves b) S waves

14. Where two oceanic plates converge, ____________ subduction zones often occur.
    a) Andean-type c) American-type e) Arctic-type
    b) Himalaya-type d) Aleutian-type

15. The San Andreas fault zone separates two great sections of Earth's crust, the North American plate and the ____________ plate.
    a) Nazca c) Pacific e) South American
    b) Juan de Fuca d) Philippine

16. Dense rocks like ____________ are thought to make up the mantle and provide the lava for oceanic eruptions.
    a) limestone c) peridotite e) rhyolite
    b) granite d) sandstone

17. In areas where unconsolidated materials are saturated with water, earthquakes can turn stable soil into a fluid during a phenomenon called ____________.
    a) libation c) leaching e) localization
    b) lithification d) liquefaction

18. It is assumed that many of the terranes found in the North American Cordillera were once crustal fragments scattered throughout the eastern ____________ ocean basin.
    a) Pacific b) Atlantic c) Indian

19. The adjustments of materials that follow a major earthquake often generate smaller earthquakes called ____________.
    a) tremors c) foreshocks e) body waves
    b) aftershocks d) surface waves

20. Faults where the movement is primarily vertical are called ____________ faults.
    a) transform c) dip-slip e) strike-slip
    b) oblique d) random

21. An earthquake with a magnitude of 6.5 releases ____________ times more energy than one with a magnitude of 5.5.
    a) 10 c) 30 e) 50
    b) 20 d) 40

22. Where oceanic crust is being thrust beneath a continental mass, ____________ subduction zones often occur.
    a) Andean-type c) American-type e) Arctic-type
    b) Himalaya-type d) Aleutian-type

23. Dip-slip faults are classified as ____________ faults when the hanging wall moves up relative to the footwall.
    a) normal
    b) transform
    c) reverse
    d) tensional
    e) strike-slip

24. Earthquake epicenters are most closely correlated with ____________.
    a) continental interiors
    b) plate boundaries
    c) population centers
    d) continental shelves
    e) high latitudes

25. Which one of the following mountain ranges has formed where continental crusts have converged?
    a) Sierra Nevada
    b) Andes Mountains
    c) Himalaya Mountains
    d) Coast Ranges

True/false. For the following true/false questions, if a statement is not completely true, mark it false. For each false statement, change the ***italicized*** *word to correct the statement.*

1. ___ Earthquake waves that travel through Earth's interior are called *body* waves.
2. ___ The adjustments that follow a major earthquake often generate smaller earthquakes called *foreshocks.*
3. ___ Earthquakes with a Richter magnitude less than *eight* are usually not felt by humans.
4. ___ Most of our knowledge of Earth's interior comes from the study of *earthquakes.*
5. ___ Vibrations known as earthquakes occur as rock slips and *elastically* returns to its original shape.
6. ___ Strike-slip faults that are associated with plate boundaries are called *transform* faults.
7. ___ Earthquake *body* waves are divided into two types called primary (P) waves and secondary (S) waves.
8. ___ Most tsumanis result from *horizontal* displacement of the ocean floor during an earthquake.
9. ___ The study of earthquakes is called *seismography.*
10. ___ No reliable method of *short-range* earthquake prediction has yet been devised.
11. ___ Fluids (gases and liquids) *cannot* transmit P waves.
12. ___ P waves arrive at a recording station *after* S waves.
13. ___ The boundary that separates the crust from the underlying mantle is known as the *shadow* discontinuity.
14. ___ Most earthquakes occur along faults associated with *plate* boundaries.

15. ___ The lithosphere is situated *below* the asthenosphere.

16. ___ Earth's *inner* core is a solid metallic sphere.

17. ___ The *epicenter* of an earthquake is the location on the surface directly above the focus.

18. ___ Most folds result from *compressional* stresses in the crust.

19. ___ To locate an epicenter, the distance from *three* or more different seismic stations must be known.

20. ___ The mantle is *solid* because both P and S waves travel through it.

21. ___ The farther a station is from an earthquake, the *greater* the difference in arrival times of the P and S waves.

22. ___ Earthquakes in the central and eastern United States occur *more* frequently than along plate-boundary areas.

23. ___ The continental crust is mostly made of *granitic* rocks.

24. ___ Most major episodes of mountain building have occurred along *divergent* plate boundaries.

25. ___ A refined Richter scale is used to describe earthquake *magnitude.*

Written questions

1. Describe an earthquake and the circumstances that cause it to occur.

2. Describe the composition (mineral makeup) of Earth's crust, mantle, and core.

3. Describe the process responsible for the formation of Earth's major mountain systems.

7

FIRE WITHIN:
Igneous Activity

Fire Within: Igneous Activity begins with a description of the catastrophic 1980 eruption of Mount St. Helens. A discussion of volcanism and the factors that determine the nature of volcanic eruptions (magma composition, temperature, and amount of dissolved gases) is followed by an examination of the materials that can be extruded during an eruption. The types of volcanic cones, their origins, shapes, and compositions, as well as the nature of fissure eruptions and volcanic landforms, are also presented.

An examination of intrusive igneous activity includes the classification and description of the major intrusive igneous bodies—dikes, sills, laccoliths, and batholiths. The chapter closes with a discussion of the relations between igneous activity and plate tectonics.

Learning Objectives

After reading, studying, and discussing this chapter, you should be able to:

- List the factors that determine the violence of volcanic eruptions.
- List the materials that are extruded from volcanoes.
- Describe the major features produced by volcanic activity.
- List and describe the major intrusive igneous features.
- Describe the relation between igneous activity and plate tectonics.

Chapter Review

- The primary factors that determine the nature of volcanic eruptions include the magma's *temperature*, its *composition*, and the *amount of dissolved gases* it contains. As lava cools, it begins to congeal, and as viscosity increases, its mobility decreases. *The viscosity of magma is directly related to its silica content. Granitic* lava, with its high silica content, is very viscous and forms short, thick flows. *Basaltic* lava, with a lower silica content, is more fluid and may travel a long distance before congealing. Dissolved gases provide the force which propels molten rock from the vent of a volcano.

- The materials associated with a volcanic eruption include *lava flows* (*pahoehoe* and *aa* flows for basaltic lavas), *gases* (primarily in the form of *water vapor*), and *pyroclastic material* (pulverized rock and lava fragments blown from the volcano's vent, which include *ash*, *pumice*, *lapilli*, *cinders*, *blocks*, and *bombs*).

- *Shield cones* are broad, slightly domed volcanoes built primarily of fluid, basaltic lava. Cinder cones have very steep slopes composed of pyroclastic material. *Composite cones*, or *stratovolcanoes*, are large, nearly symmetrical structures built of interbedded lavas and pyroclastic deposits. Composite cones represent the most violent type of volcanic activity.
- Other than volcanoes, regions of volcanic activity may contain *volcanic necks* (rocks that once occupied the vents of volcanoes but are now exposed because of erosion), *craters* (steep walled depressions at the summit of most volcanoes), *calderas* (craters that exceed one kilometer in diameter), *fissure eruptions* (volcanic material extruded from fractures in the crust), and *pyroclastic flows*.
- Igneous intrusive bodies are classified according to their *shape* and by their *orientation with respect to the host rock*, generally sedimentary rock. The two general shapes are *tabular* (sheet-like) and *massive*. Intrusive igneous bodies that cut across existing sedimentary beds are said to be *discordant*, whereas those that form parallel to existing sedimentary beds are *concordant*.
- *Dikes* are tabular, discordant igneous bodies produced when magma is injected into fractures that cut across rock layers. Tabular, concordant bodies called *sills* form when magma is injected along the bedding surfaces of sedimentary rocks. *Laccoliths* are similar to sills but form from less-fluid magma that collects as a lens-shaped mass that arches the overlying strata upward. *Batholiths*, the largest intrusive igneous bodies with surface exposures of more than 100 square kilometers (40 square miles), frequently compose the cores of mountains.
- Active areas of *volcanism* are found *along the oceanic ridges*, *adjacent to ocean trenches*, as well as *the interiors of plates* themselves. Most active volcanoes are associated with plate boundaries.

Chapter Outline

I. Volcanic eruptions
    A. Factors that determine violence
        1. Composition of magma
        2. Temperature of magma
        3. Dissolved gases in magma
    B. Viscosity of magma
        1. Viscosity is resistance to flow
        2. Factors affecting viscosity
            a. Temperature—hotter—less viscous
            b. Composition (silica content)
                1. High silica—high viscosity
                2. Granitic lava is very viscous
                3. Low silica—more fluid
                4. Basaltic lava has low viscosity
            c. Dissolved gases
                1. Mainly water vapor and carbon dioxide
                2. Gases expand near surface
                3. Provide force to extrude lava
        4. Violence related to ease of escape
            a. Easy escape from fluid magma
            b. Viscous magma more violent
II. Materials associated with volcanic eruptions
    A. Lava flows
        1. Basaltic lavas more fluid
        2. Types of lava
            a. Pahoehoe lava—resembles braids in ropes
            b. Aa lava—rough, jagged blocks
    B. Gases
        1. % to 5% percent of magma by weight
        2. Mainly water vapor and carbon dioxide
    C. Pyroclasts
        1. "Fire fragments"
        2. Types of pyroclastic material

a. Ash—fine, glassy fragments
b. Pumice—from “frothy” lava
c. Lapilli—“walnut” size
d. Cinders—“pea-sized” with voids
e. Particles larger than lapilli
1. Blocks—hardened lava
2. Bombs—ejected as hot lava

D. Nuée Ardente
1. Fiery cloud
2. Hot gases infused with ash
3. Speeds up to 200 km (125 miles) per hour

III. Volcanoes
A. General features
1. Opening at summit
a. Crater—steep-walled depression at summit
b. Caldera—large summit depression greater than 1 km diameter
2. Vent—connects crater to magma chamber
B. Types of volcanoes
1. Shield volcano
a. Broad, slightly domed
b .Primarily of basaltic (fluid) lava
c. Generally large
d. Generally produce large volume of lava
e. e.g., Mauna Loa in Hawaii
2. Cinder cone
a. Built from ejected lava fragments
b. Steep slope angle
c. Rather small
d. Frequently occur in groups
3.Composite cone (or stratovolcano)
a. Most encircle Pacific Ocean (e.g., Fujiyama, Mt. Shasta)
b. Large
c. Interbedded lavas and pyroclastics
d. Most violent type of activity (e.g., Vesuvius)

IV.Volcanic landforms
A. Volcanic neck
1. Resistant vent left standing after erosion
2. e.g., Ship Rock, New Mexico
B. Crater or caldera
C. Fissure eruption and lava plateau
1. Volcanic material extruded from fractures
2. e.g., Columbia Plateau
D. Pyroclastic flow
1. From silica-rich magma
2. Consists of ash and pumice fragments
3. Propelled from vent at high speed
4. e.g., Yellowstone plateau

V. Intrusive igneous activity
A. Magma emplaced at depth
B. Underground igneous body is called a pluton
C. Plutons are classified according to
1. Shape
a. Tabular (sheetlike)
b. Massive
2. Orientation with respect to host rock
a. Discordant—cut across sedimentary beds
b. Concordant—parallel to sedimentary beds
D. Types of igneous intrusive features
1. Dike—tabular, discordant pluton
2. Sill
a. Tabular, concordant pluton
b. e.g., Palisades Sill, NY
3. Laccolith
a. Forms in same way as sill
b. Lens shaped mass
c. Arches overlying strata upward
4. Batholith
a. Largest intrusive body
b. Surface exposure 100 square km plus—smaller bodies are termed stocks
c. Frequently form cores of mountains

VI. Igneous activity and plate tectonics
A. Distribution of igneous activity is not random
1. Granitic or andesitic rocks are found on the continents and continental margins
2. Volcanoes in deep-ocean basins extrude basaltic lavas

B. Global patterns of igneous activity
  1. Igneous activity along plate margins
    a. Oceanic ridge spreading center
      1. Lithosphere pulls apart
      2. Less pressure on underlying rocks
      3. Partial melting
      4. Large quantities of basaltic magma
    b. Convergent plate margin
      1. Subduction zone (trench)
      2. Descending plate partially melts
      3 .Magma slowly rises upward
      4. Rising magma can form
        a. Island arc in ocean
        b. Andesitic-granitic volcanoes on continent
      5. Associated with Pacific Basin
        a. Called "Ring of Fire"
        b. Explosive—high gas volcanoes
  2. Intraplate volcanism
    a. Activity within a rigid plate
    b. Basaltic magma source
      1. Partial melting of mantle rock
      2. Plumes of hot mantle material
        a. Form hot spots on surface
        b. One may be below Hawaii
    c. Granitic magma source
      1. Remelting of continental crust
      2. Situated over hot mantle plume

Vocabulary Review

Choosing from the list of key terms, furnish the most appropriate response for the following statements.

1. Particles of pulverized rock, lava, and glass fragments blown from the vent of a volcano are referred to as ____________.

2. A(n) ____________ is an unusually large volcanic summit depression that exceeds one kilometer in diameter.

3. A lava flow that has a surface of rough, jagged blocks is called a(n) ____________.

4. Ship Rock, New Mexico, a feature produced by erosion, is an example of a(n) ____________.

5. A(n) ____________ is a sheetlike body that is produced when magma is injected into a fracture that cuts across rock layers.

6. The crater of a volcano is connected to a magma chamber via a pipelike conduit called a(n) ____________.

7. A(n) ____________ is a rather small volcano with steep slopes built from ejected lava fragments.

8. By far the largest intrusive igneous body is a(n) ____________.

9. A volcano that takes the shape of a broad, slightly domed structure is called a(n) ____________.

Key Terms

Page numbers shown in ( ) refer to the textbook page where the term first appears.

aa flow (p. 183)
batholith (p. 195)
caldera (p. 184)
cinder cone (p. 186)
composite cone (stratovolcano) (p. 187)
crater (p. 184)
dike (p. 193)
fissure (p. 191)
flood basalt (p. 192)
hot spot (p. 198)
laccolith (p. 195)
mantle plume (p. 198)
nuée ardente (p. 188)
pahoehoe flow (p. 183)
pyroclastic flow (p. 192)
pycroclast (p. 184)
shield volcano (p. 185)
sill (p. 193)
vent (p. 184)
viscosity (p. 180)
volcanic neck (p. 191)
volcano (p. 184)

10. A flow of fluid basaltic lava that issues from cracks or fissures and commonly covers an extensive area to a thickness of hundreds of meters is called a(n) ____________.

11. A lava flow with a smooth-to-ropy appearance that is produced from fluid basaltic lava is called a(n) ____________.

12. A hot plume, which may extend to Earth's core-mantle boundary, produces a(n) ____________, a volcanic region a few hundred kilometers across.

13. A(n) ____________ is a large, nearly symmetrical volcano built of interbedded strata of lavas and pyroclastic deposits.

14. A(n) ____________ is a mountainous accumulation of material formed by successive eruptions from a central vent.

15. A(n) ____________ is an igneous intrusive feature that forms from a lens-shaped mass of magma that arches the overlying strata upward.

16. A narrow fracture, break, or crack in the crust is known as a(n) ____________.

17. A(n) ____________ is a tabular pluton formed when magma is injected along sedimentary bedding surfaces.

18. Hot gases infused with incandescent ash ejected from a volcano produce a fiery cloud called a(n) ____________ that flows down steep slopes at high speed.

19. Most igneous rocks melt over a temperature range of a few hundred degrees, a process known as ____________, which produces most, if not all, magma.

20. A material's ____________ is a measure of its resistance to flow.

21. The source of some intraplate basaltic magma are hot __________ that are thought to originate at great depth.

Comprehensive Review

1. List three factors that determine whether a volcano extrudes magma violently or "gently."

   1)

   2)

   3)

2. Using Figure 7.1, select the letter of the diagram that illustrates each of the following types of volcanoes. Name a volcano that is an example of each type. Also describe the eruptive pattern most commonly associated with each type.

   a) Shield volcano: ____ (Example:)

   b) Cinder cone: ____ (Example:)

   c) Composite cone (stratovolcano): ____ (Example:)

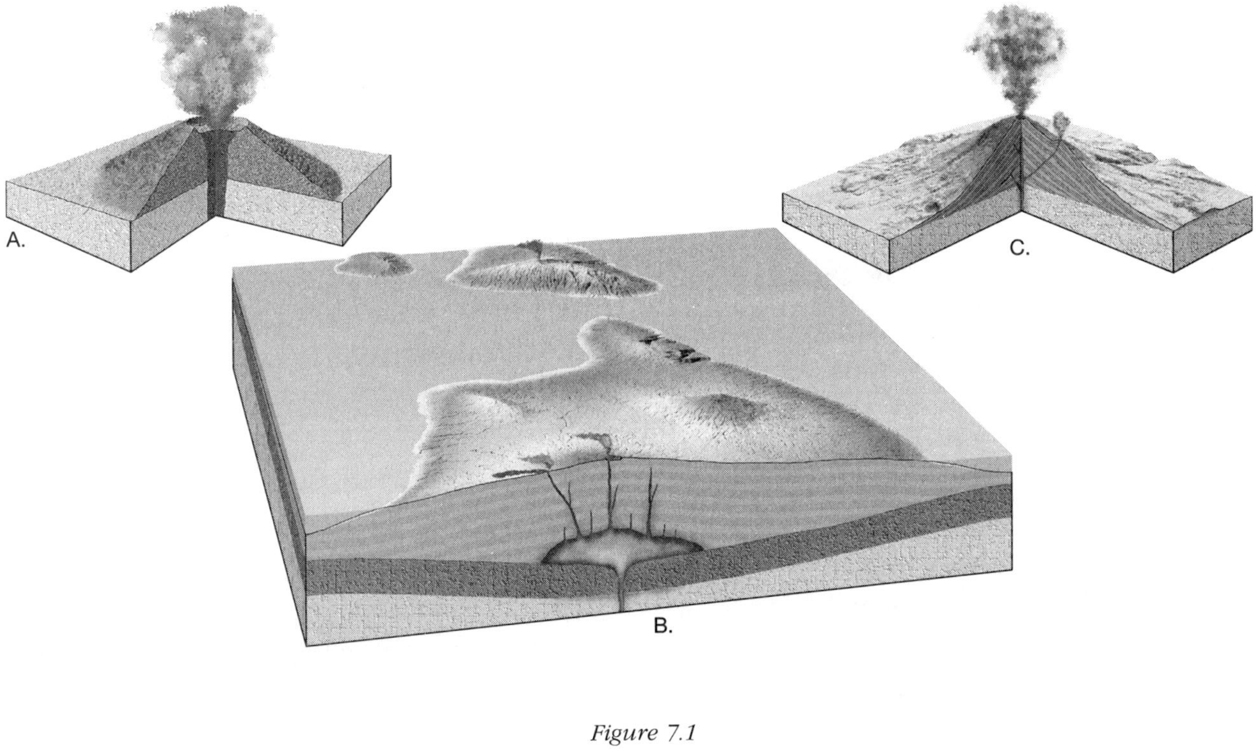

Figure 7.1

3. What are the major gases released during a volcanic eruption?

4. Describe the meaning of the following terms as they are related to igneous plutons.
   a) Discordant:

   b) Concordant:

5. What factors affect the viscosity of magma?

6. Using Figure 7.2, select the letter that illustrates each of the following igneous intrusive features.
   a) Sill: ____
   b) Batholith: ____
   c) Laccolith: ____
   d) Dike: ____

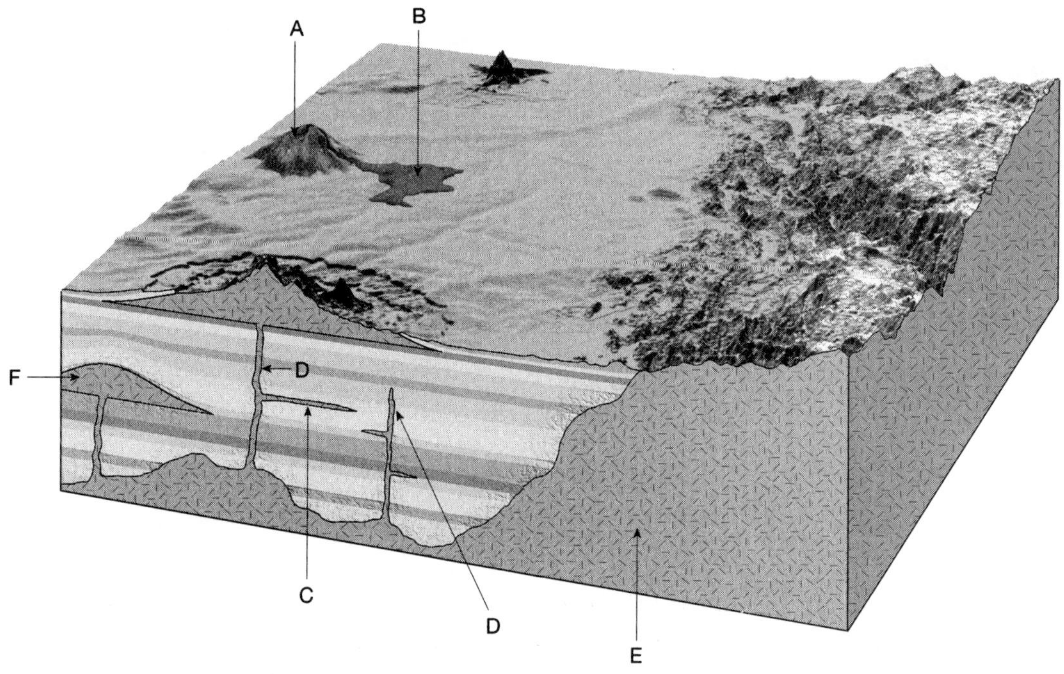

Figure 7.2

7. Compare basaltic magma to granitic magma in terms of
   a) Silica content:

   b) Viscosity:

8. List the three major zones of volcanic activity and relate each to global tectonics.
   1)

   2)

   3)

Practice Test

Multiple choice. Choose the best answer for the following multiple choice questions.

1. Underground igneous rock bodies are called __________.
   a) aquifers c) playas e) placers
   b) plutons d) pluvials

2. The greatest volume of volcanic material is produced by __________.
   a) cinder cones c) laccoliths e) explosive eruptions
   b) fissure eruptions d) shield cones

3. Highly viscous magmas tend to impede the upward migration of expanding gases, which often results in __________ eruptions.
   a) relatively quiet b) explosive

4. The most violent type of volcanic activity is associated with __________.
   a) cinder cones c) composite cones e) shield cones
   b) sills d) intermediate cones

5. Which one of the following is NOT a factor that determines the violence of a volcanic eruption?
   a) temperature of the magma c) the magma's composition
   b) size of the volcanic cone d) amount of dissolved gases in the magma

6. The most abundant gas produced during Hawaiian eruptions is __________.
   a) nitrogen c) oxygen e) water vapor
   b) carbon dioxide d) chlorine

7. The force that extrudes magma from a volcanic vent is provided by ____________.
   a) dissolved gases c) the magma's heat e) discordant plutons
   b) gravity d) the volcano's slope

8. The area of igneous activity commonly called the Ring of Fire surrounds the ____________.
   a) Indian Ocean c) Atlantic Ocean e) Pacific Ocean
   b) Coral Sea d) Sea of Japan

9. A magma's viscosity is directly related to its ____________.
   a) depth c) volcanic cone e) color
   b) age d) silica content

10. Pulverized rock, lava, and glass fragments produced from the vent of a volcano are known as ____________.
    a) nuée ardentes c) pyroclasts e) pahoehoes
    b) sills d) craters

11. Which type of volcano consists of interbedded strata of lavas and pyroclastic material?
    a) cinder cone c) intermediate cone e) pyro-cone
    b) composite cone d) shield cone

12. Fluid basaltic lavas of the Hawaiian type commonly form ____________.
    a) aa flows c) pyroclastic flows e) lapilli
    b) pahoehoe flows d) nuée ardente

13. Basaltic lava tends to be ____________ fluid than granitic lava.
    a) more b) less

14. When silica-rich magma is extruded, ash and pumice fragments may be propelled from the vent at high speeds and produce ____________.
    a) flood basalts c) batholiths e) pyroclastic flows
    b) pahoehoe flows d) a shield volcano

15. Unusually large volcanic summit depressions that exceed one kilometer in diameter are known as ____________.
    a) calderas c) craters e) laccoliths
    b) vents d) sills

16. Large particles of hardened lava ejected from a volcano are termed ____________.
    a) bombs c) welded tuff e) blocks
    b) lapilli d) cinders

17. Eruptions of fluid basaltic lavas, such as those that occur in Hawaii, tend to be ____________.
    a) relatively quiet b) unpredictable c) extremely violent d) explosive

18. Hot gases infused with incandescent ash ejected from a volcano often produce a fiery cloud called a(n) ____________.
    a) nuée ardente c) lapilli e) volcanic bomb
    b) laccolith d) pyroclastic

19. The type of volcano produced almost entirely of pyroclastic material is the ___________.
    a) shield volcano   c) composite cone   e) batholith
    b) cinder cone   d) pyro-cone

20. A(n) ___________ is a tabular, concordant pluton.
    a) dike   c) stock   e) batholith
    b) laccolith   d) sill

21. Which type of volcanoes are generally small and occur in groups?
    a) composite cones   c) shield cones   e) intermediate cones
    b) laccoliths   d) cinder cones

22. The largest intrusive igneous bodies are ___________.
    a) dikes   c) batholiths   e) sills
    b) stocks   d) laccoliths

23. Intraplate volcanism may be associated with the formation of ___________ over rising plumes of hot mantle material.
    a) hot spots   c) subduction zones   e) volcanic necks
    b) dikesd   d) ocean ridges

True/false. For the following true/false questions, if a statement is not completely true, mark it false. For each false statement, change the ***italicized*** *word to correct the statement.*

1. ___ Most of Earth's active volcanoes are near *divergent* plate margins.
2. ___ The more viscous a material, the *greater* its resistance to flow.
3. ___ Plutons similar to but smaller than batholiths are termed *stocks.*
4. ___ Water vapor is the *most* abundant gas in magma.
5. ___ The smallest volcanoes are *composite* cones.
6. ___ The greatest volume of volcanic rock is produced along the oceanic *ridge* system.
7. ___ Magmas that produce basaltic rocks contain *more* silica than those that form granitic rocks.
8. ___ Located at the summit of most volcanoes is a steep-walled depression called a *sill.*
9. ___ *Dikes* are tabular, discordant plutons.
10. ___ With increasing depth in Earth's interior, there is a gradual *decrease* in temperature.
11. ___ A magma's viscosity is directly related to its *iron* content.
12. ___ Most of the volcanoes of the Cascade Range in the northwestern United States are *shield* cones.

13. ___ When *subduction* volcanism occurs in the ocean, a chain of volcanoes called an island arc is produced.

14. ___ The large expanse of granitic rock exposed in the interior of North America is called the *Canadian* Shield.

15. ___ The viscosity of magma, plus the quantity of dissolved *gases* and the ease with which they can escape, determines the nature of volcanic eruptions.

16. ___ One of the best known *batholiths* in North America is along the Hudson River near New York.

17. ___ Secondary volcanic vents that emit only gases are called *fumaroles.*

18. ___ The Columbia River Plateau in the northwestern United States formed from very fluid *basaltic* lava that erupted from numerous fissures.

Written questions

1. What is the difference between magma and lava?

2. List the three main types of volcanoes and describe the shape of each.

3. List and describe three igneous intrusive features.

8

Geologic Time

Geologic Time opens with a brief history of geology that spans from James Ussher to James Hutton and Sir Charles Lyell. The chapter continues with a discussion of the fundamental principles of relative dating: law of superposition, principle of original horizontality, principle of cross-cutting relationships, and the uses of inclusions and unconformities. How rock units in different localities can be correlated is also investigated. The significance of fossils to understanding geologic time begins with a discussion of the conditions favoring preservation. Also examined is the use of fossils in correlating and dating rock units. Following an explanation of radioactivity, the fundamentals of radiometric dating are presented. The chapter concludes with an examination of the geologic time scale.

Learning Objectives

After reading, studying, and discussing this chapter, you should be able to:

- Describe the doctrine of uniformitarianism.
- Explain the difference between absolute and relative dating.
- List the laws and principles used in relative dating.
- Discuss unconformities.
- Explain correlation of rock layers.
- Describe fossils, fossilization, and the uses of fossils.
- Explain radioactivity and radiometric dating.
- Describe the geologic time scale.

Chapter Review

- The *doctrine of uniformitarianism*, one of the fundamental principles of modern geology put forth by James Hutton in the late 1700s, states that the physical, chemical, and biological laws that operate today have also operated in the geologic past. The idea is often summarized as, "the present is the key to the past." Hutton argued that processes that appear to be slow-acting could, over long spans of time, produce effects that were just as great as those resulting from sudden catastrophic events. *Catastrophism*, on the other hand, states that Earth's landscapes have been developed primarily by great catastrophes.
- The two types of dates used by geologists to interpret Earth history are 1) *relative dates*, which put events in their *proper sequence of formation*, and 2) *absolute dates*, which pinpoint the *time in years* when an event took place.
- Relative dates can be established using the *law of superposition*, *principle of original horizontality*, *principle of cross-cutting relationships*, *inclusions*, and *unconformities*.

• *Correlation*, the matching up of two or more geologic phenomena in different areas, is used to develop a geologic time scale that applies to the whole Earth.

• *Fossils* are the remains or traces of prehistoric life. The special conditions that favor preservation are *rapid burial* and the possession of *hard parts* such as shells, bones, or teeth.

• Fossils are used to correlate sedimentary rocks that are from different regions by using the rocks' distinctive fossil content and applying the *principle of fossil succession.* The principle of fossil succession states that fossil organisms succeed one another in a definite and determinable order, and therefore any time period can be recognized by its fossil content.

• *Radioactivity* is the spontaneous breaking apart (decay) of certain unstable atomic nuclei. Three common forms of radioactive decay are 1) emission of alpha particles from the nucleus, 2) emission of a beta particle (or electron) from the nucleus, and 3) capture of an electron by the nucleus.

• An unstable *radioactive isotope*, called the *parent*, will decay and form *daughter products.* The length of time for one-half of the nuclei of a radioactive isotope to decay is called the *half-life* of the isotope. If the half-life of the isotope is known, and the parent/daughter ratio can be measured, the age of a sample can be calculated.

• The *geologic time* scale divides Earth's history into units of varying magnitude. It is commonly presented in chart form, with the oldest time and event at the bottom and the youngest at the top. The principle subdivisions of the geologic time scale, called *eons*, include the *Hadean*, *Archean*, *Proterozoic* (together, these three eons are commonly referred to as the *Precambrian*), and, beginning about 570 million years ago, the *Phanerozoic.* The Phanerozoic (meaning "visible life") eon is divided into the following *eras: Paleozoic* ("ancient life"), *Mesozoic* ("middle life"), and *Cenozoic* ("recent life").

• A significant problem in assigning absolute dates to units of time is that *not all rocks can be dated radiometrically.* A sedimentary rock may contain particles of many ages that have been weathered from different rocks that formed at various times. One way geologists assign absolute dates to sedimentary rocks is to relate them to datable igneous masses, such as volcanic ash beds.

Chapter Outline

I. Historical notes
- A. Catastrophism
  - 1. Landscape developed by catastrophes
  - 2. James Ussher, mid-1600s, concluded that the Earth was only a few thousand years old
- B. Modern geology
  - 1. Uniformitarianism
    - a. Fundamental principle of geology
    - b. "The present is the key to the past"
  - 2. James Hutton
    - a. *Theory of the Earth*
    - b. Late 1700s
  - 3. Sir Charles Lyell
    - a. Advanced modern geology
    - b. *Principles of Geology*
    - c. Mid-1800s

II. Relative dating
- A. Placing rocks and events in sequence
- B. Principles and rules of
  - 1. Law of superposition—oldest rocks on bottom
  - 2. Principle of original horizontality—sediment deposited horizontally
  - 3. Principle of cross-cutting relationships—younger feature cuts through older
  - 4. Inclusions
    - a. One rock contained within another
    - b. Rock containing inclusions is younger
  - 5. Unconformities
    - a. An unconformity is a break in the rock record

b. Types of unconformities
  1. Angular unconformity—tilted rocks overlain by flat-lying rocks
  2. Disconformity—strata on either side are parallel
  3. Nonconformity
     a. Metamorphic or igneous rocks below
     b. Younger sedimentary rocks above

III. Correlation of rock layers
  A. Matching rocks of similar age
  B. Often relies upon fossils

IV. Fossils
  A. Remains or traces of prehistoric life
  B. Conditions favoring preservation
     1. Rapid burial
     2. Possession of hard parts
  C. Fossils and correlation
     1. Principle of fossil succession
        a. Fossils succeed one another in order
        b. William Smith—late 1700s and early 1800s
     2. Index fossils
        a. Widespread
        b. Short range of geologic time

V. Radioactivity and radiometric dating
  A. Radioactivity
     1. Breaking apart of atomic nuclei
     2. Atomic number—number of protons in an atom's nucleus
     3. Mass number
        a. Protons plus neutrons in an atom's nucleus
        b. Isotope
           1. Number of neutrons varies from parent atom
           2. Has different mass number than parent atom
     4. Radioactive decay
        a. Parent—unstable isotope
        b. Daughter products—isotopes from decay of parent
        c. Methods of radioactive decay
           1. Alpha emission
           2. Beta emission
           3. Electron capture
  B. Radiometric dating
     1. Half-life—time for one-half of the nuclei to decay
     2. Requires a closed system
     3. Cross-checks used for accuracy
     4. Complex procedure
     5. Yields absolute dates

VI. Geologic time scale
  A. Divides geologic history into units
  B. Originally created using relative dates
  C. Subdivisions
     1. Eon
        a. Greatest expanse of time
        b. Names
           1. Phanerozoic ("visible life")—most recent
           2. Proterozoic
           3. Archean
           4. Hadean—oldest
     2. Era
        a. Subdivision of an eon
        b. Eras of the Phanerozoic eon
           1. Cenozoic ("recent life")
           2. Mesozoic ("middle life")
           3. Paleozoic ("ancient life")
     3. Eras are subdivided into periods
     4. Periods are subdivided into epochs
  D. Difficulties in dating the time scale
     1. Not all rocks are datable
     2. Sedimentary ages are rarely reliable
     3. Materials often used to bracket events

Vocabulary Review

Choosing from the list of key terms, furnish the most appropriate response for the following statements.

1. Simply stated, the doctrine of ____________ states that the physical, chemical, and biological laws that operate today have also operated in the geologic past.

2. The ____________ is the age of "recent life" on Earth.

Key Terms

Page numbers shown in ( ) refer to the textbook page where the term first appears.

absolute date (p. 206)
angular unconformity (p. 209)
catastrophism (p. 205)
Cenozoic era (p. 218)
conformable (p. 208)
correlation (p. 210)
cross-cutting relationships, principle of (p. 208)
disconformity (p. 209)
eon (p. 217)
epoch (p. 218)
era (p. 218)
fossil (p. 210)
fossil succession, principle of (p. 212)
geologic time scale (p. 217)
half-life (p. 215)
inclusions (p. 208)
index fossil (p. 214)
Mesozoic era (p. 218)
nonconformity (p. 209)
original horizontality, principle of (p. 206)
Paleozoic era (p. 218)
period (p. 218)
Phanerozoic eon (p. 218)
Precambrian (p. 220)
radioactivity (p. 215)
radiometric dating (p. 215)
relative dating (p. 206)
superposition, law of (p. 206)
unconformity (p. 209)
uniformitarianism (p. 205)

3. To state specifically the point in history when something took place, for example, the extinction of the dinosaurs about 66 million years ago, is to use a type of date called a(n) ____________.

4. The "matching-up" of rocks from similar ages but different regions is referred to as ____________.

5. The time required for one-half of the nuclei in a radioactive sample to decay is called the ____________ of the isotope.

6. The doctrine of ____________ adheres to the idea that Earth's landscape was produced by sudden and often worldwide disasters of unknowable causes that no longer operate.

7. To place events in their proper sequence is to apply a type of dating technique called ____________.

8. The Hadean, the Archean, and the Proterozoic eons are commonly referred to as the ____________.

9. To assume that rock layers that are inclined have been moved into that position by crustal disturbances is to apply a principle of relative dating known as ____________.

10. To state that "in an undeformed sequence of sedimentary rocks the oldest rock is at the bottom" is to use a basic principle of relative dating called the ____________.

11. A(n) ____________ is a break in the rock record during which deposition ceased, erosion removed previously formed rocks, and then deposition resumed.

12. A(n) ____________ is the remains or trace of prehistoric life.

13. The spontaneous breaking apart of atomic nuclei is a process known as ____________.

14. A(n) ____________ is a subdivision of a geologic era.

15. Pieces of one rock unit contained within another are called ____________.

16. The unit of geologic time known as the ____________ is noted for its "ancient life."

17. A fossil of an organism that was widespread geographically but limited to a short span of geologic time is often referred to as a(n) ____________.

18. The fact that fossils succeed one another in a definite and determinable order is known as the ____________.

19. Using radioactive isotopes to calculate the ages of rocks and minerals is a procedure called ____________.

20. The unit of geologic time called a(n) ____________ represents the greatest expanse of time.

Comprehensive Review

1. Write a brief statement that compares the doctrine of catastrophism with the doctrine of uniformitarianism.

2. Dinosaurs became extinct about 66 million years ago, while large coal swamps flourished in North America about 300 million years ago. Using these facts, write relative and absolute date statements.

   a) Relative date:

   b) Absolute date:

3. Briefly state the following:

   a) Law of superposition:

   b) Principle of original horizontality:

   c) Principle of fossil succession:

4. Using Figure 8.1, select the letter of the diagram that illustrates each of the following types of unconformities. (Unconformities are indicated with a dark wavy line and the small "v" symbol represents igneous rock.)

   a) Nonconformity: ____

   b) Angular unconformity: ____

   c) Disconformity: ____

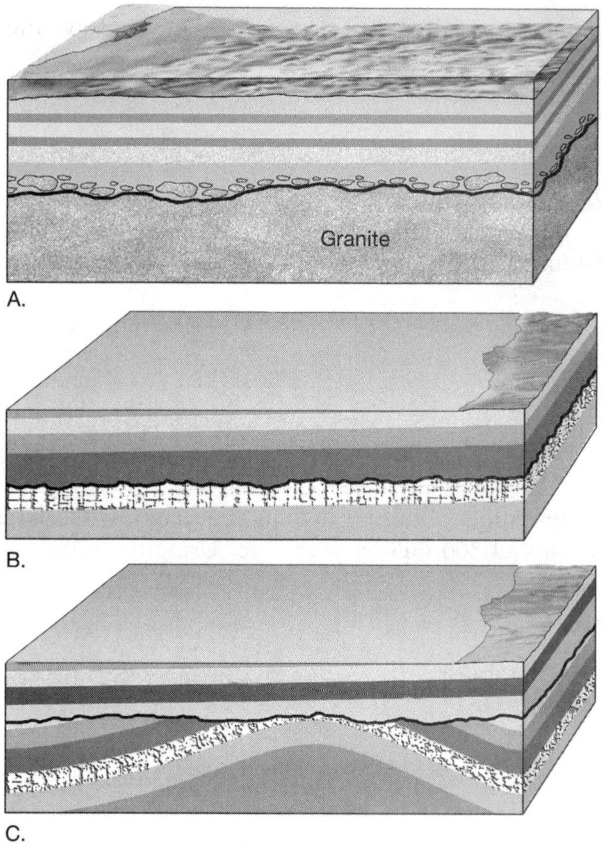

Figure 8.1

5. List the two special conditions that are necessary for an organism to be preserved as a fossil.

   1)

   2)

6. Examine the illustrations of the sedimentary rocks exposed along two separate cliffs in Figure 8.2. Different types of rocks are represented with different patterns. Write the letter of the rock layer at location 2 that correlates with the indicated rock layer at location 1.

	Location 1		**Location 2**
a)	layer b	correlates with	layer ____
b)	layer d	correlates with	layer ____

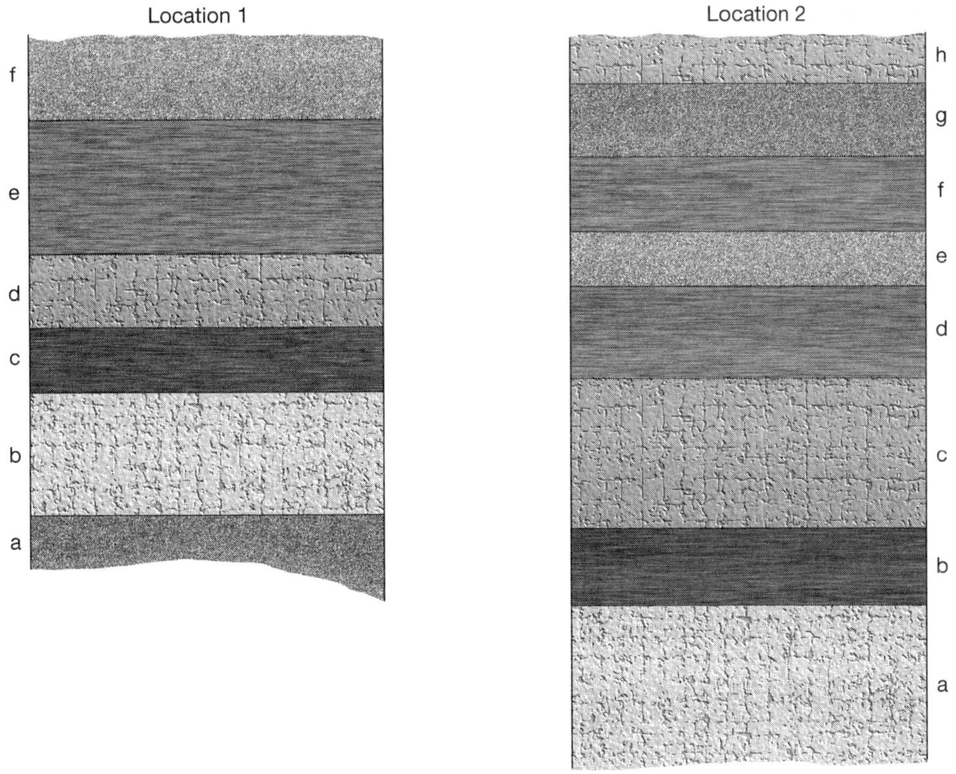

Figure 8.2

7. Explain the difference between an atom's atomic number and its mass number.

8. List three common types of radioactive decay.
   1)

   2)

   3)

9. Why can radiometric dating be considered reliable?

10. Briefly summarize the events of the following eras of the Phanerozoic eon.
    a) Cenozoic era:

    b) Mesozoic era:

    c) Paleozoic era:

11. What is the primary problem in assigning absolute dates to the units of time of the geologic time scale?

12. Using Figure 8.3, answer the following questions concerning the relative ages of the features illustrated. Also, when indicated, list the law or principle of relative dating you used to arrive at your answer.
    a) Dike B is ___________ [older, younger] than fault B.

       Law or principle:

    b) The shale is ___________ [older, younger] than the sandstone.

       Law or principle:

    c) Dike B is ___________ [older, younger] then the batholith.

    d) The sandstone is ___________ [older, younger] than Dike A.

    e) The conglomerate is ___________[older, younger] than the shale.

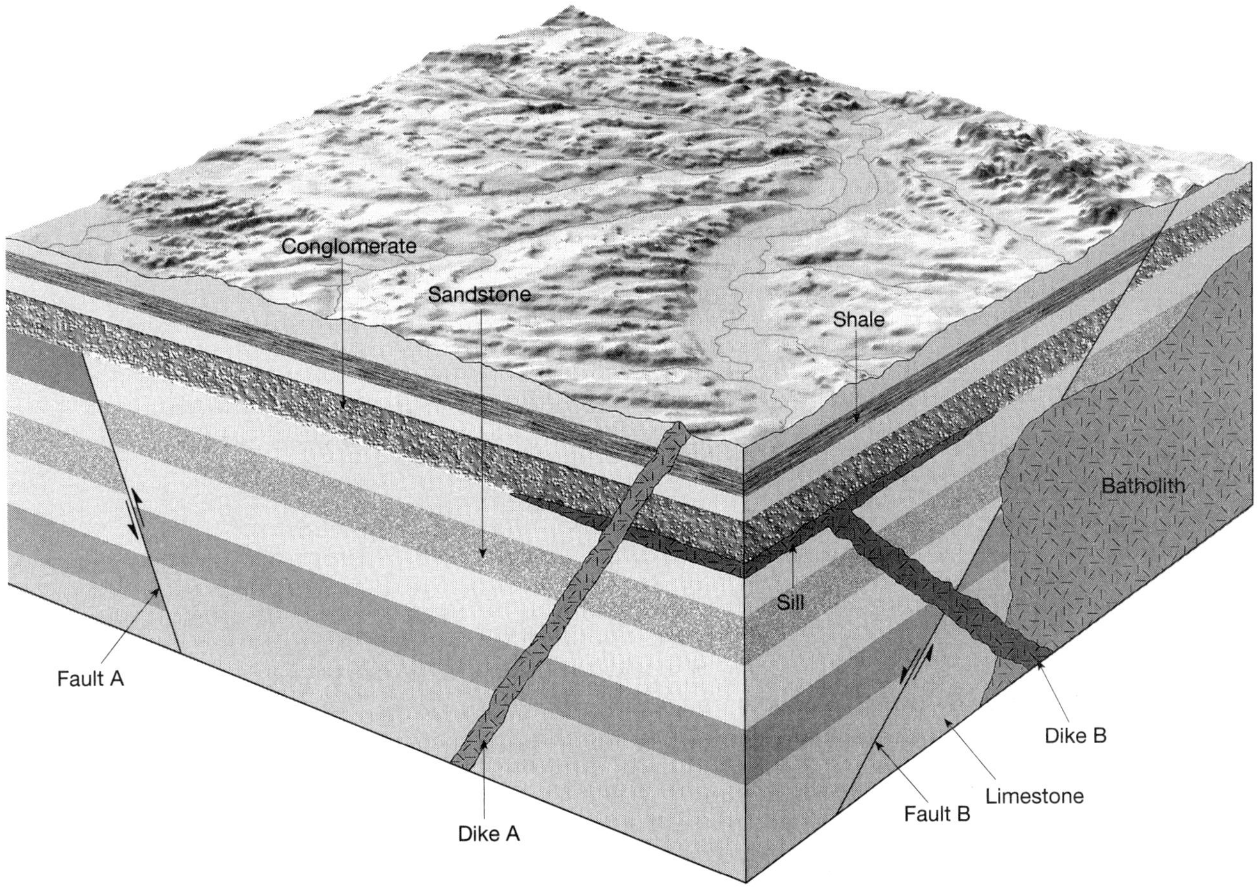

Figure 8.3

13. Assume a radioactive isotope has a half-life of 50,000 years. If the ratio of radioactive parent to stable daughter product is 1:3, how old is the rock containing the radioactive material?

Practice Test

Multiple choice. Choose the best answer for the following multiple choice questions.

1. "The physical, chemical, and biological laws that operate today have also operated in the geologic past," is a statement of the doctrine of ____________.

   a) uniformitarianism c) catastrophism e) paleontology
   b) universal truth d) unity

2. The task of matching up rocks of similar ages in different regions is known as ____________.

   a) indexing c) linking e) superposition
   b) correlation d) succession

3. Fossils that are widespread geographically and are limited to a short span of time are referred to as ____________ fossils.

   a) key c) succeeding e) relative
   b) matching d) index

4. A(n) ____________ in the rock record represents a long period during which deposition ceased, erosion removed previously formed rocks, and then deposition resumed.

   a) absolute date c) relative date e) unconformity
   b) superposition d) inclusion

5. The process by which atomic nuclei spontaneously break apart (decay) is termed ____________.

   a) ionization c) reduction e) radioactivity
   b) fusion d) nucleation

6. The doctrine of ____________ implies that Earth's landscape has been developed by worldwide disasters over a short time span.

   a) uniformitarianism c) catastrophism e) suddenness
   b) destruction d) universal time

7. The greatest expanses of geologic time are referred to as ____________.

   a) eras c) eons e) systems
   b) periods d) epochs

8. Nicolaus Steno proposed the most basic principle of relative dating, the law of ____________.

   a) superposition c) sequential stacking e) suddenness
   b) secondary intrusion d) succession

9. Which era of the geologic time scale is often referred to as the "age of mammals"?

   a) Cenozoic b) Mesozoic c) Paleozoic

10. The doctrine of ____________ was put forth by James Hutton in his monumental work *Theory of the Earth.*

    a) destruction c) suddenness e) universal time
    b) catastrophism d) uniformitarianism

11. Rock layers that have been deposited essentially without interruption are said to be ____________ strata.

    a) included c) abnormal e) sequential
    b) conformable d) succeeding

12. Which era of the geologic time scale is often referred to as the "age of reptiles"?
    a) Cenozoic b) Mesozoic c) Paleozoic

13. The type of date that places events in proper sequence is referred to as a(n) ___________ date.
    a) sequential c) relative e) temporary
    b) secondary d) positional

14. The remains or traces of prehistoric life are known as ___________.
    a) indicators c) replicas e) fissures
    b) fossils d) paleotites

15. Between 1830 and 1872, ___________ produced eleven editions of his great work *Principles of Geology*.
    a) Archbishop James Ussher d) William Smith
    b) James Hutton e) Charles Darwin
    c) Sir Charles Lyell

16. Atoms with the same atomic number but different mass numbers are called ___________.
    a) protons c) ions e) nucleoids
    b) isotopes d) variants

17. A break that separates older metamorphic rocks from younger sedimentary rocks immediately above them is a type of unconformity called a(n) ___________.
    a) disconformity c) angular unconformity e) pseudoconformity
    b) nonconformity d) contact unconformity

18. "Most layers of sediments are deposited in a horizontal position" is a statement of the principle of ___________.
    a) original horizontality c) fossil succession e) cross-bedding
    b) cross-cutting relationships d) sediment

19. Which one of the following is NOT an era of the Phanerozoic eon?
    a) Proterozoic b) Mesozoic c) Paleozoic d) Cenozoic

20. Pieces of one rock unit contained within another are called ___________.
    a) intrusions c) hosts e) inclusions
    b) interbeds d) conformities

21. Which one of the following is NOT a common type of radioactive decay?
    a) alpha particle emission c) beta particle emission
    b) nuclei ionization d) electron capture

22. Earth is about ___________ years old.
    a) 4,000 c) 5.8 million e) 6.4 billion
    b) 4.6 million d) 4.6 billion

23. A date that pinpoints the time in history when something took place is known as a(n) ___________ date.
    a) relative c) known e) factual
    b) exact d) absolute

24. During radioactive decay, what will be the parent/daughter ratio after three half-lives?
    a) 1:3 c) 1:5 e) 1:7
    b) 1:4 d) 1:6

25. The principle of fossil __________ states that fossil organisms succeed one another in a definite and determinable order.
    a) inclusions c) sequences e) evolution
    b) succession d) superposition

True/false. For the following true/false questions, if a statement is not completely true, mark it false. For each false statement, change the ***italicized*** *word to correct the statement.*

1. ___ The entire geologic time scale was created using *absolute* dates.
2. ___ Acceptance of the concept of uniformitarianism means the acceptance of a very *short* history for Earth.
3. ___ *Index* fossils are useful for correlating the rocks of one area with those of another.
4. ___ To state that a rock is 120 million years old is an example of applying a(n) *absolute* date.
5. ___ Rapid burial and the possession of *soft* parts are necessary conditions for fossilization.
6. ___ The rates of geologic processes have undoubtedly *varied* through time.
7. ___ All rocks can be *dated* radiometrically.
8. ___ Every element has a different number of *electrons* in the nucleus.
9. ___ *Catastrophists* believed that Earth was only a few thousand years old.
10. ___ When magma works its way into a rock and crystallizes, we can assume that the intrusion is *older* than the rock that has been intruded.
11. ___ During radioactive decay, as the percentage of radioactive parent atoms declines, the proportion of stable daughter atoms *decreases.*
12. ___ In addition to being important tools for correlation, fossils are also useful *environmental* indicators.
13. ___ Most layers of sediment are deposited in a *horizontal* position.
14. ___ Widespread occurrence of the first organisms with shells marks the beginning of the *Phanerozoic* eon.
15. ___ The fossil record of those organisms with *hard* body parts that lived in areas of sedimentation is quite abundant.
16. ___ No place on Earth has a complete set of *conformable* strata.

17. ___ Among the major contributions of geology is the concept that Earth history is exceedingly *long.*

18. ___ The "present is the key to the past" summarizes the basic idea of *catastrophism.*

19. ___ *Index* fossils are important time indicators.

20. ___ An accurate radiometric date can only be obtained if the material remains in a *closed* system during the entire period since its formation.

Written questions

1. Explain the difference between radiometric and relative dating.

2. List three laws or principles that are used to determine relative dates.

3. How are fossils helpful in geologic investigations?

9 OCEANS: *The Last Frontier*

Oceans: The Last Frontier introduces oceanography with a brief discussion of the extent and distribution of the world ocean. After examining the composition of seawater, the ocean's layered temperature and salinity structures are presented. The ocean floor section of this chapter begins with an examination of continental margins and continues with submarine canyons and associated turbidity currents, ocean basin features, and mid-ocean ridges. The origin of coral reefs and atolls is also presented. The chapter ends with a discussion of sea floor sediments and how they are used to study climatic changes.

Learning Objectives

After reading, studying, and discussing this chapter, you should be able to:

- Describe the extent and boundaries of the world ocean.
- Discuss the chemical composition of ocean water.
- Explain the ocean's layered temperature and salinity structures.
- Describe the major features of the continental margin, ocean basin floor, and mid-ocean ridges.
- List the types of sea floor sediments.
- Describe how ocean floor sediments relate to climatic changes.

Chapter Review

- *Oceanography* is a composite science that draws on the methods and knowledge of biology, chemistry, physics, and geology to study all aspects of the world ocean.
- *Earth is a planet dominated by oceans.* Seventy-one percent of Earth's area consists of oceans and marginal seas. In the Southern Hemisphere, often called the *water hemisphere*, about 81% of the surface is water. Of the three major oceans, Pacific, Atlantic, and Indian, the Pacific Ocean is the largest, contains slightly *more than half of the water* in the world ocean, and has the *greatest average depth*–3940 meters (12,900 feet).
- *Salinity* is the proportion of dissolved salts to pure water, usually expressed in parts per thousand (‰). The average salinity in the open ocean ranges from 35‰ to 37‰. The principal elements that contribute to the ocean's salinity are *chlorine* (55%) and *sodium* (31%). The primary *sources for the salts* in the ocean are chemical weathering of rocks on the continents and *outgassing* through volcanism. Outgassing is also considered to be the principal source of water in the oceans as well as in the atmosphere.

• In most regions, open oceans exhibit a *three-layered temperature and salinity structure.* Ocean water temperatures are warmest at the surface because of solar energy. The mixing of waves as well as the turbulence from currents can distribute this heat to a depth of about 450 meters or more. Beneath the sun-warmed zone of mixing, a layer of rapid temperature change, called the *thermocline*, occurs. Below the thermocline, in the deep zone, temperatures fall only a few more degrees. The changes in salinity with increasing depth correspond to the general three-layered temperature structure. In the low and middle latitudes, a surface zone of higher salinity is underlain by a layer of rapidly decreasing salinity, called the *halocline.* Below the halocline, salinity changes are small.

• The zones that collectively make up the *continental margin* include the *continental shelf* (a gently sloping, submerged surface extending from the shoreline toward the deep-ocean basin), *continental slope* (the true edge of the continent, which has a steep slope that leads from the continental shelf into deep water), and in regions where trenches do not exist, the steep continental slope merges into a gradual incline known as the *continental rise.* The continental rise consists of sediments that have moved downslope from the continental shelf to the deep-ocean floor.

• *Submarine canyons* are deep, steep-sided valleys that originate on the continental slope and may extend to depths of three kilometers. Many submarine canyons have been excavated by *turbidity currents* (downslope movements of dense, sediment-laden water).

• The *ocean basin floor* lies between the continental margin and the mid-oceanic ridge system. The features of the ocean basin floor include *deep-ocean trenches* (the deepest parts of the ocean, where moving crustal plates descend into the mantle), *abyssal plains* (the most level places on Earth, consisting of thick accumulations of sediments that were deposited atop the low, rough portions of the ocean floor by turbidity currents), and *seamounts* (isolated volcanic peaks on the ocean floor that originate near oceanic ridges or in association with volcanic hot spots).

• *Mid-ocean ridges,* the sites of sea-floor spreading, are found in all major oceans and represent more than 20 percent of Earth's surface. These broad features are characterized by an elevated position, extensive faulting, and volcanic structures that have developed on newly formed oceanic crust. Most of the geologic activity associated with ridges occurs along a narrow region on the ridge crest, called the *rift zone,* where magma from the asthenosphere moves upward to create new slivers of oceanic crust.

• *Coral reefs,* which are confined largely to the warm, sunlit waters of the Pacific and Indian Oceans, are constructed over thousands of years primarily from the skeletal remains and secretions of corals and certain algae. Coral islands, called *atolls,* consist of a continuous or broken ring of coral reef surrounding a central lagoon. Atolls form from corals that grow on the flanks of sinking volcanic islands, where the corals continue to build the reef complex upward as the island sinks.

• *There are three broad categories of sea floor sediments. Terrigenous sediment* consists primarily of mineral grains that were weathered from continental rocks and transported to the ocean. *Biogenous sediment* consists of shells and skeletons of marine animals and plants. *Hydrogenous sediment* includes minerals that crystallize directly from seawater through various chemical reactions. Sea floor sediments are helpful when studying worldwide climate changes because they often contain the remains of organisms that once lived near the sea surface. The numbers and types of these organisms change as the climate changes, and their remains in the sediments record these changes.

Chapter Outline

I. The world ocean
   A. Global extent of the world ocean
      1. 71% of Earth
      2. 61% of Northern Hemisphere
      3. 81% of Southern Hemisphere
   B. Size and depth of the major oceans
      1. Pacific by far the largest
      2. Pacific has greatest average depth
      3. Atlantic the shallowest
   C. Composition of seawater
      1. 3.5% (by weight) dissolved minerals
      2. Salinity
         a. Proportion of dissolved salts
         b. Measured in parts-per-thousand (‰)
         c. Salinity of open ocean: 33‰–37‰
         d. Major constituents: Cl^- and Na^+
      3. Sources of salts
         a. Chemical weathering of rocks
         b. Outgassing
   D. The ocean's three-layered structure
      1. Temperature
         a. Warmest at surface
         b. Thermocline–rapid decrease with depth
         c. Below thermocline–little change
      2. Salinity
         a. In low and middle latitudes
            1. Higher at surface
            2. Fresh water is evaporated
         b. Halocline–rapid decrease with depth
         c. Below halocline–little change

II. Earth beneath the sea
   A. Measuring depth
      1. Echo sounder
         a. Primary instrument for measuring depth
         b. Reflects sound from ocean floor
   B. Major topographic units of the sea floor
      1. Continental margin
         a. Continental shelf
            1. Flooded extension of continent
            2. Shoreline toward ocean basin
            3. Gently sloping
            4. Submerged surface
         b. Continental slope
            1. Seaward edge of shelf
            2. Steep gradient into deep water
            3. True edge of continent
            4. Associated with continental slopes are
               a. Turbidity currents–dense, sediment-laden water
               b. Submarine canyons
               c. Continental rise
                  1. Found where trenches are absent
                  2. Thick accumulation of sediment
                  3. At base of slope
                  4. Turbidity currents form–deep-sea fans
      2. Ocean basin floor
         a. Deep-ocean trench
            1. Deepest part of ocean
            2. Where plates plunge into mantle
         b. Abyssal plain
            1. Most level place on Earth
            2. Thick accumulations of sediment
            3. Found in all oceans
         c. Seamount
            1. Isolated volcanic peak
            2. Some form over hot spot
            3. Mid-ocean ridge
               a. Site of sea-floor spreading
               b. Found in all major oceans
               c. Rift zone
                  1. On ridge crest
                  2. Geologically active
                  3. Magma moves up from asthenosphere
                  4. New ocean crust forms here
   C. Coral reef and atoll
      1. Coral reef
         a. Built up over thousands of years by
            1. Remains and secretions of corals
            2. Certain algae
         b. Thrive in warm waters
      2. Atoll
         a. Coral reef surrounding a lagoon
         b. Formation explained by Darwin
         c. Forms on flanks of sinking volcanic island
   D. Sea floor sediment
      1. Thickness varies
      2. Mud is the most common sediment
      3. Types of sediments

a. Terrigenous sediment–weathered continental rocks
b. Biogenous sediment
   1. Shells, skeletons, and plants
   2. Calcareous and siliceous oozes
c. Hydrogenous sediment
   1. Chemicals direct from seawater
   2. Manganese nodules
4. Climatic change
   a. Life in the sea changes with climate
   b. Remains of organisms record changes

Vocabulary Review

Choosing from the list of key terms, furnish the most appropriate response for the following statements.

1. ____________ is the composite science that draws on the methods and knowledge of biology, chemistry, physics, and geology to study all aspects of the world ocean.

2. The proportion of dissolved salts to pure water, expressed in parts-per-thousand, is referred to as ____________.

3. A long, narrow, deep ocean trough is known as a(n) ____________.

4. The ____________ is the gently sloping submerged surface of the continental margin that extends from the shoreline toward the deep-ocean basin.

5. The downslope movement of dense, sediment laden water is known as a(n) ____________.

6. Likely one of the most level places on Earth, a(n) ____________ is an incredibly flat feature on the ocean floor.

7. The ____________ is the site of sea-floor spreading.

8. The principal source of water in the oceans as well as in the atmosphere is a process known as ____________.

9. Sediment that consists primarily of mineral grains that were weathered from continental rocks and transported to the ocean is known as ____________.

10. A(n) ____________ consists of a continuous or broken ring of coral reef surrounding a central lagoon.

Key Terms

Page numbers shown in ( ) refer to the textbook page where the term first appears.

abyssal plain (p. 237)
atoll (p. 238)
biogenous sediment (p. 240)
continental margin (p. 235)
continental rise (p. 235)
continental shelf (p. 235)
continental slope (p. 235)
coral reef (p. 238)
deep-ocean trench (p. 236)
deep-sea fan (p. 236)
echo sounder (p. 231)
guyot (p. 237)
halocline (p. 231)
hydrogenous sediment (p. 240)
mid-ocean ridge (p. 237)
oceanography (p. 228)
outgassing (p. 230)
rift zone (p. 238)
salinity (p. 229)
seamount (p. 237)
submarine canyon (p. 236)
terrigenous sediment (p. 239)
thermocline (p. 231)
turbidity current (p. 236)

11. The ____________ is the ocean layer where there is a rapid change in salinity.

12. A(n) ____________ is constructed primarily from the skeletal remains and secretions of corals and certain algae.

13. An isolated volcanic peak found on the ocean floor is known as a(n) ____________.

14. Sediment consisting of minerals that crystallize directly from seawater through various chemical reactions is called ____________.

15. The layer of the ocean where there is a rapid temperature change is known as the ____________.

16. Ocean sediment consisting of shells and skeletons of marine animals and plants is known as ____________.

17. In regions where trenches do not exist, the steep continental slope merges into a more gradual incline known as the ____________.

18. A(n) ____________ is a submerged, flat-topped seamount.

19. Much of the geologic activity on the ocean floor occurs along a relatively narrow zone on the mid-ocean ridge crest known as the ____________.

20. The ____________ marks the seaward edge of the continental shelf.

21. The instrument used to measure ocean depths by transmitting sound waves toward the ocean floor is the ____________.

Comprehensive Review

1. Briefly contrast the distribution of land and water in the Northern Hemisphere with that in the Southern Hemisphere.

2. Describe the three-layered structure normally found in the open ocean.

3. What are the principal elements that contribute to the ocean's salinity?

4. Using Figure 9.1, select the letter that identifies each of the following features of the continental margin.

   a) Deep-sea fan: ____
   b) Submarine canyon: ____
   c) Continental slope: ____
   d) Continental shelf:____
   e) Continental rise:____

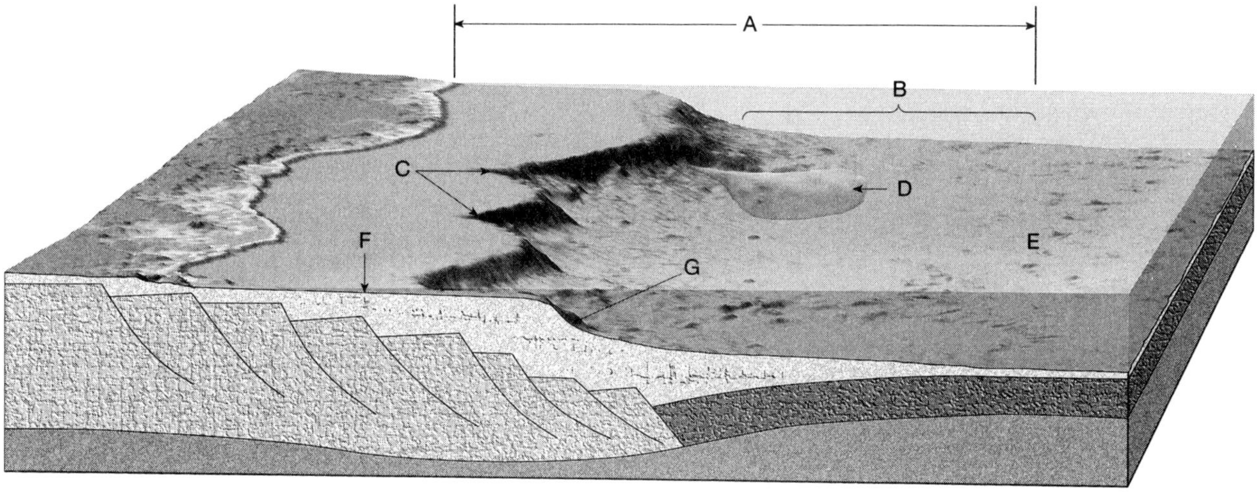

Figure 9.1

5. Describe abyssal plains and explain their origin. In Figure 9.1, letter ____ illustrates an abyssal plain.

6. What are the two primary sources for the salts in the ocean?
   1)

   2)

7. Explain the reason for the difference in the ocean's surface salinity between subtropical and equatorial regions.

8. What is the probable origin for submarine canyons that are not seaward extensions of river valleys?

9. List three water conditions that are necessary for the growth of coral.

   1)

   2)

   3)

10 Briefly describe the formation of an atoll.

11 List the three broad categories of sea floor sediments and briefly describe the origin of each.

   1)

   2)

   3)

12. Using Figure 9.2, select the letter that identifies each of the following ocean basin features.

   a) Deep-ocean trench: ____ d) Rift zone: ____

   b) Seamount: ____ e) Guyot: ____

   c) Mid-ocean ridge: ____

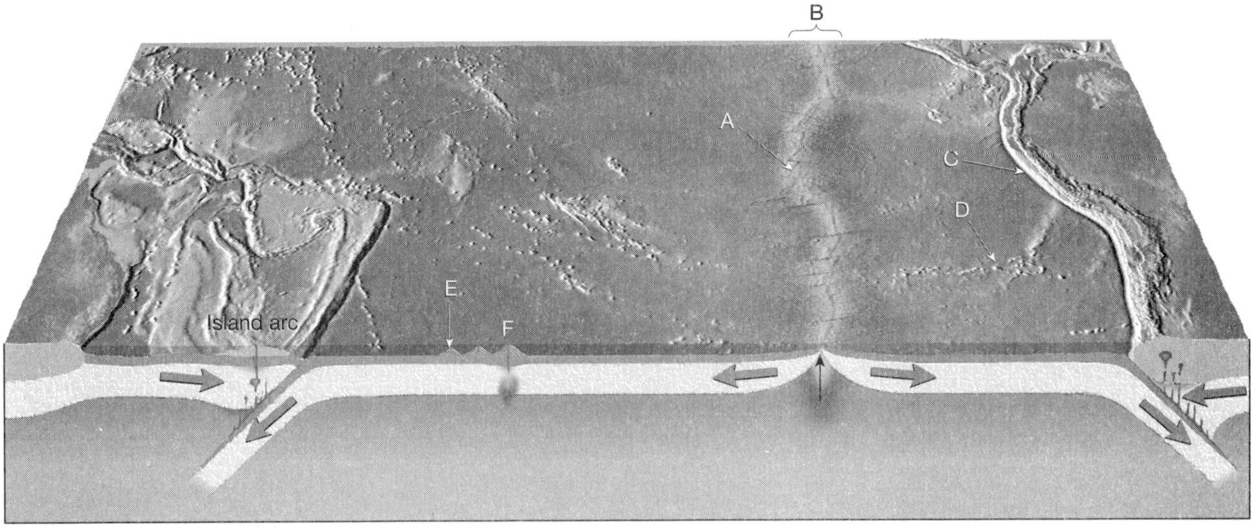

Figure 9.2

Practice Test

Multiple choice. Choose the best answer for the following multiple choice questions.

1 .Approximately ____________ percent of Earth's area is represented by oceans and marginal seas.

a) 50 c) 70 e) 90
b) 60 d) 80

2. Important mineral deposits, including large reservoirs of petroleum and natural gas, are associated with ____________.

a) continental shelves b) ocean trenches c) mid-ocean ridges d) rift zones

3. Seawater consists of about ____________ percent (by weight) of dissolved mineral substances.

a) 1.5 c) 3.5 e) 5.5
b) 2.5 d) 4.5

4. The most prominent feature in the oceans, forming an almost continuous mountain range, are the ____________.

a) deep-ocean trenches c) seamounts e) guyots
b) mid-ocean ridgesd d) abyssal plains

5. The layer of rapid temperature change in the ocean is known as the ____________.

a) halocline c) thermocline e) tempegrad
b) centicline d) faricline

6. The largest of Earth's water bodies is the ____________.

a) Pacific Ocean c) Atlantic Ocean e) Indian Ocean
b) Mediterranean Sea d) Gulf of Mexico

7. The principal element that contributes to the ocean's salinity is ____________.

a) sodium c) bromine e) chlorine
b) potassium d) calcium

8. As a consequence of the distribution of land and water on Earth, the Southern Hemisphere is referred to as the ____________ hemisphere.

a) water c) land e) dry
b) open d) mixed

9. Which of the following is NOT one of the three major topographic units of the ocean basins?

a) continental margins b) ocean basin floor c) coastal plain d) mid-ocean ridges

10. The seaward edge of the continental shelf is marked by the ____________.

a) abyssal plain c) continental slope e) shoreline
b) ocean trench d) mid-ocean ridge

11. Which process is thought to be the principal source of water in the oceans as well as in the atmosphere?

a) sublimation c) deposition e) freezing
b) outgassing d) crystallization

12. Sediment that consists primarily of mineral grains that were weathered from continental rocks and transported to the ocean is known as ____________ sediment.
    a) biogenous b) hydrogenous c) terrigenous

13. The layer of rapid salinity change in the ocean is known as the ____________.
    a) halocline c) thermocline e) halograd
    b) saligrad d) salicline

14. The ____________ are likely the most level places on Earth.
    a) mid-ocean ridges c) deep-ocean trenches e) guyots
    b) abyssal plains d) continental slopes

15. Which one of the following is NOT a zone included in the continental margin?
    a) continental slope c) continental coastal plain
    b) continental rise d) continental shelf

16. Which of the following is NOT considered one of the three major oceans?
    a) Atlantic Ocean b) Arctic Ocean c) Pacific Ocean d) Indian Ocean

17. The fact that elements such as chlorine and bromine are more abundant in the ocean than in Earth's crust confirms the hypothesis that ____________ is largely responsible for the present oceans.
    a) subduction c) dissolving e) precipitation
    b) convergence d) volcanism

18. Salinity variations in the open ocean normally range from 33‰ to ____________.
    a) 37‰ c) 60‰ e) 83‰
    b) 45‰ d) 67‰

19. Depths in the ocean are often measured using a(n) ____________.
    a) altimeter c) submersible e) echo-sounder
    b) laser d) rope

20. Which one of the following water bodies has the highest salinity?
    a) Lake Michigan c) Red Sea e) Hudson Bay
    b) Baltic Sea d) Gulf of Mexico

21. Deep, steep-sided valleys that originate on the continental slope and may extend to depths of 3 km are referred to as ____________.
    a) slope canyons c) rifts e) submarine canyons
    b) abyssal plains d) trenches

22. The most abundant salt in the sea is ____________.
    a) calcium chloride c) potassium chloride e) sodium fluoride
    b) magnesium chloride d) sodium chloride

23. Where trenches do not exist, the steep continental slope merges into a more gradual incline known as the continental ____________.
    a) abyss c) end e) run
    b) rise d) coast

24. The deepest parts of the ocean are long, narrow features known as deep-ocean ____________.
    a) ridges c) scars e) rifts
    b) trenches d) holes

25. Isolated volcanic peaks on the ocean floor are known as ____________.
    a) ridges c) nodules e) rifts
    b) abyssals d) seamounts

26. Downslope movements of dense, sediment-laden water are known as ____________ currents.
    a) avalanche c) density e) longshore
    b) turbidity d) sediment

27. Which one of the following is NOT a water condition necessary for coral growth?
    a) warm b) sunlit c) deep d) clear

28. The speed of sound in water is about ____________ meters per second.
    a) 500 c) 1500 e) 2500
    b) 1000 d) 2000

True/false. For the following true/false questions, if a statement is not completely true, mark it false. For each false statement, change the ***italicized*** *word to correct the statement.*

1. ___ Ocean temperature and salinity vary with depth and generally follow a *three-layered* structure in the open ocean.
2. ___ Because it has many shallow adjacent seas, the *Pacific* Ocean is the shallowest of the three major oceans.
3. ___ The true edge of the continent is the continental *slope.*
4. ___ In the ocean, *high* salinities are found where evaporation is high.
5. ___ Turbidity currents originate along the continental *rise.*
6. ___ *Low* ocean surface salinities occur where large quantities of fresh water are supplied by rivers.
7. ___ Deep-ocean *trenches* in the open ocean are often paralleled by volcanic island arcs.
8. ___ In the ocean where heavy precipitation occurs, *higher* surface salinities prevail.
9. ___ The most common sediment covering the deep-ocean floor is *sand.*
10. ___ Below the thermocline, temperatures *fall* only a few more degrees.
11. ___ Much of the geologic activity along mid-ocean ridges occurs in the *rift* zone.
12. ___ The continental *shelf* varies greatly in width from one continent to another.
13. ___ Deep-ocean trenches are the sites where moving crustal plates plunge back into the *mantle.*

14. ___ Mid-ocean ridges exist in *all* major oceans.

15. ___ The Northern Hemisphere can be referred to as the *land* hemisphere.

16. ___ Atolls owe their existence to the *sudden* sinking of oceanic crust.

Written questions

1. Describe the term salinity. What is the average salinity of the ocean?

2. What are the three major topographic units of the ocean basins?

3. Relate deep-ocean-trenches and mid-ocean ridges to sea-floor spreading.

10

The Restless Ocean

The Restless Ocean opens with a discussion of ocean surface currents and their importance. Deep-ocean circulation is also briefly examined. Tides are analyzed, along with a detailed look at wave mechanics and wave erosion. Also investigated are the shoreline features and erosional problems that are caused by wave action and rising sea levels. The chapter concludes with a look at emergent and submergent coastal regions.

Learning Objectives

After reading, studying, and discussing this chapter, you should be able to:

- List the factors that influence surface ocean currents.
- Discuss the importance of surface ocean currents.
- Describe deep-ocean circulation.
- Discuss the factors that influence tides.
- Describe wave characteristics and types.
- Describe wave erosion and the features produced by wave erosion.
- Discuss shoreline erosional problems and solutions.
- Explain the differences between an emergent and submergent coast.

Chapter Review

- *Surface ocean currents* are parts of huge, slowly moving, circular whirls, or *gyres*, that begin near the equator in each ocean. *Wind is the driving force* for the ocean's surface currents. Where wind is in contact with the ocean, it passes energy to the water through friction and causes the surface layer to move. The most significant factor other than wind that influences the movement of ocean waters is the *Coriolis effect*, the deflective force of Earth's rotation which causes free-moving objects to be deflected to the right in the Northern Hemisphere and to the left in the Southern Hemisphere. Because of the Coriolis effect, surface currents form clockwise gyres in the Northern Hemisphere and counterclockwise gyres in the Southern Hemisphere.
- Ocean currents are important in navigation and travel and for the effect that they have on climates.The moderating effect of poleward-moving warm ocean currents during the winter in middle latitudes is well known.
- In contrast to surface currents, *deep-ocean circulation* is governed by gravity and driven by *density* differences. The two factors that are most significant in creating a dense mass of water are *temperature* and *salinity*.

• *Tides*, the daily rise and fall in the elevation of the ocean surface at a specific location, are caused by the *gravitational attraction* of the moon and, to a lesser extent, by the sun. Near the times of new and full moons, the sun and moon are aligned and their gravitational forces are added together to produce especially high and low tides. These are called the *spring tides.* Conversely, at about the times of the first and third quarters of the moon, when the gravitational forces of the moon and sun are at right angles, the daily *tidal range* is less. These are called *neap tides.*

• The three factors that influence the *height*, *wavelength*, and *period* of a wave are 1) *wind speed*, 2) *length of time the wind has blown*, and 3) *fetch*, the distance that the wind has traveled across the open water.

• The two types of wind-generated waves are 1) *waves of oscillation*, which are waves in the open sea in which the wave form advances as the water particles move in circular orbits, and 2) *waves of translation*, the turbulent advance of water formed near the shore as waves of oscillation collapse, or *break*, and form *surf.*

• Features produced by *shoreline erosion* include *wave-cut cliffs* (which originate from the cutting action of the surf against the base of coastal land), *wave-cut platforms* (relatively flat, benchlike surfaces left behind by receding cliffs), *sea arches* (formed when a headland is eroded and two caves from opposite sides unite), and *sea stacks* (formed when the roof of a sea arch collapses).

• Some of the features formed when sediment is moved by *beach drift* and *longshore currents* are *spits* (elongated ridges of sand that project from the land into the mouth of an adjacent bay), *baymouth bars* (sand bars that completely cross a bay), and *tombolos* (ridges of sand that connect an island to the mainland or to another island).

• Local factors that influence shoreline erosion are 1) the proximity of a coast to sediment-laden rivers, 2) the degree of tectonic activity, 3) the topography and composition of the land, 4) prevailing winds and weather patterns, and 5) the configuration of the coastline and nearshore areas.

• Three basic responses to shoreline erosion problems are 1) building *structures* such as *groins* (short walls built at a right angle to the shore to trap moving sand) and *seawalls* (barriers constructed to prevent waves from reaching the area behind the wall) to hold the shoreline in place, 2) *beach nourishment*, which involves the addition of sand to replenish eroding beaches, and 3) *relocate* buildings away from the beach.

• One frequently used classification of coasts is based upon changes that have occurred with respect to sea level. *Emergent coasts*, often with wave-cut cliffs and wave-cut platforms above sea level, develop either because an area experiences uplift or as a result of a drop in sea level. Conversely, *submergent coasts*, with their drowned river mouths, called *estuaries*, are created when sea level rises or the land adjacent to the sea subsides.

Chapter Outline

I. Ocean water movements
   A. Surface currents
      1. Huge, slowly moving gyres
      2. Generated by wind
      3. Related to atmospheric circulation
      4. Deflected by the Coriolis effect
         a. To right in Northern Hemisphere
         b. To left in Southern Hemisphere

5. Importance of surface currents
   a. Navigation
   b. Influence climates

B. Upwelling
   1. Vertical water movement
   2. Along eastern shores of oceans
   3. Caused by
      a. Coriolis effect
      b. Surface water moving from shore

C. Deep-water circulation
   1. Governed by gravity
   2. Driven by density differences
   3. Factors creating a dense mass of water
      a. Temperature—cold water is dense
      b. Salinity—density increases with increasing salinity
   4. Called thermohaline circulation
   5. Dense water masses are created in
      a. Arctic regions
      b. Antarctic regions

D. Tides
   1. Changes in elevation of ocean surface
   2. Caused by gravitational forces of the
      a. Moon, and to a lesser extent the
      b. Sun
   3. Types of tides
      a. Spring tide
         1. During new and full moons
         2. Gravitational forces added together
         3. Especially high and low tides
         4. Large daily tidal range
      b. Neap tide
         1. First and third quarters of the moon
         2. Gravitational forces are offset
         3. Daily tidal range is least
   4. Other influencing factors
      a. Shape of coastline
      b. Configuration of ocean basin
   5. Tidal currents
      a. Horizontal flow accompanying tides
      b. Types of tidal currents
         1. Flood current—advances into coastal zone
         2. Ebb current—seaward-moving water

E. Wind-generated waves
   1. Derive energy and motion from wind
   2. Parts
      a. Crest
      b. Trough
   3. Measurements of a wave
      a. Wave height—distance between trough and crest
      b. Wavelength—horizontal distance between crests
      c. Wave period—time interval between crests
   4. Height, length, and period depend upon
      a. Wind speed
      b. Length of time wind blows
      c. Fetch—distance wind travels
   5. Types of waves
      a. Wave of oscillation
         1. In open sea
         2. Shape moves forward
      b. Wave of translation
         1. Wave breaks along shore
         2. Water advances up the shore
         3. Forms surf
   6. Wave erosion is
      a. Wave impact and pressure
      b. Abrasion by rock fragments
   7. Wave refraction
      a. Bending of a wave along a coast
      b. Wave arrives parallel to shore
      c. Results
         1. Wave energy directed against headland
         2. Wave erosion straightens an irregular shoreline
   8. Moving sand along the beach
      a. Beach drift—sediment moves in zigzag pattern
      b. Longshore current
         1. Current in surf zone
         2. Flows parallel to coast

II. Shoreline features
   A. Features caused by wave erosion
      1. Wave-cut cliff
      2. Wave-cut platform
      3. Associated with headlands

a. Sea arch
b. Sea stack
B. Related to beach drift and longshore currents
1. Spit
a. Ridge of sand from land into bay
b. End often hooks landward
2. Baymouth bar—sand bar that completely crosses a bay
3. Tombolo—connects island to mainland
C. Barrier island
1. Mainly along Atlantic and Gulf coasts
2. Parallels the coast
3. 3 to 30 kilometers offshore
4. Originates in several ways
D. Result of shoreline erosion and deposition eventually produce a straighter coast
III. Shoreline erosion problems
A. Influenced by the local factors
1. Proximity to sediment-laden rivers
2. Degree of tectonic activity
3. Topography and composition of land
4. Prevailing wind and weather patterns
5. Configuration of the coastline
B. Three basic responses to erosion problems
1. The building of structures
a. Groin
1. Barrier at right angle to beach
2. Traps sand
b. Seawall
1. Barrier parallel to shore
2. Stops waves from reaching shore
2. Often not effective
3. Addition of sand to replenish beaches
a. Called beach nourishment
b. Not a permanent solution
4. Relocate buildings away from beach
IV. Emergent and submergent coasts
A. Emergent coast
1. Caused by
a. Uplift of an area, or
b. A drop in sea level
2. Features of an emergent coast
a. Wave-cut cliffs
b. Wave-cut platforms
B. Submergent coast
1. Caused by
a. Land adjacent to sea subsides, or
b. Sea level rises
2. Features of a submergent coast
a. Highly irregular shoreline
b. Estuaries—drowned river mouths

Vocabulary Review

Choosing from the list of key terms, furnish the most appropriate response for the following statements.

1. An elongated ridge of sand that projects from the land into the mouth of an adjacent bay is called a(n) ____________.
2. ____________ is the movement of sediment in a zigzag pattern along a beach.
3. The large circular surface current pattern found in each ocean is call a(n) ____________.
4. The ____________ is the vertical distance between the trough and crest of a wave.
5. Deep-ocean circulation is called ____________.
6. Turbulent water created by breaking waves is called ____________.
7. The deflective force of Earth's rotation on all free-moving objects is known as the ____________.
8. A wave in open water is called a(n) ____________.
9. The horizontal distance separating successive wave crests is the ____________.

10. A(n) ___________ is a low ridge of sand that parallels a coast at a distance from 3 to 30 kilometers offshore.

11. The horizontal flow of water accompanying the rise and fall of a tide is called a(n) ___________.

Key Terms

Page numbers shown in ( ) refer to the textbook page where the term first appears.

abrasion (p. 256)	submergent coast (p. 264)
barrier island (p. 260)	surf (p. 256)
baymouth bar (p. 260)	thermohaline circulation (p. 250)
beach drift (p. 257)	tidal current (p. 253)
beach nourishment (p. 263)	tidal flat (p. 253)
Coriolis effect (p. 248)	tide (p. 251)
emergent coast (p. 277)	tombolo (p. 260)
estuary (p. 265)	upwelling (p. 249)
fetch (p. 254)	wave-cut cliff (p. 259)
groin (p. 261)	wave-cut platform (p. 259)
gyre (p. 248)	wave height (p. 254)
longshore current (p. 257)	wavelength (p. 254)
sea arch (p. 259)	wave of oscillation (p. 255)
sea stack (p. 259)	wave of translation (p. 256)
seawall (p. 263)	wave period (p. 254)
spit (p. 260)	wave refraction (p. 256)

12. When a wave of oscillation breaks along a shore it becomes a type of wave called a(n) ___________.

13. The rising of colder water from deeper ocean layers to replace warmer surface water is called ___________.

14. A(n) ___________ is a ridge of sand that connects an island either to the mainland or to another island.

15. The time in terval between the successive passage of wave crests is known as the ___________.

16. When a sea arch collapses it leaves an isolated remnant called a(n) ___________.

17. A coast that develops either because the area experiences uplift or a drop in sea level is a(n) ___________.

18. A(n) ___________ is a drowned river mouth.

19. The area affected by alternating tidal currents is known as a(n) ___________.

20. The bending of a wave is called ___________.

21. ___________ is the term that describes the distance that wind has traveled across the open water.

22. A near-shore current that flows parallel to the shore is called a(n) ___________.

23. A(n) ___________ is a relatively flat, benchlike surface left behind by a receding wave-cut cliff.

24. The ___________ is the daily change in the elevation of the ocean surface.

25. A(n) ___________ is a barrier built at a right angle to the beach to trap sand that is moving parallel to the shore.

Comprehensive Review

1. Briefly describe the influence that the Coriolis effect has on the movement of ocean waters.

2. Using Figure 10.1, select the letter that identifies each of the following surface currents.

a) Gulf Stream: ____

b) Peruvian: ____

c) Canaries: ____

d) West Wind Drift: ____

e) Benguela: ____

f) Labrador: ____

g) North Pacific Drift: ____

h) Equatorial: ____

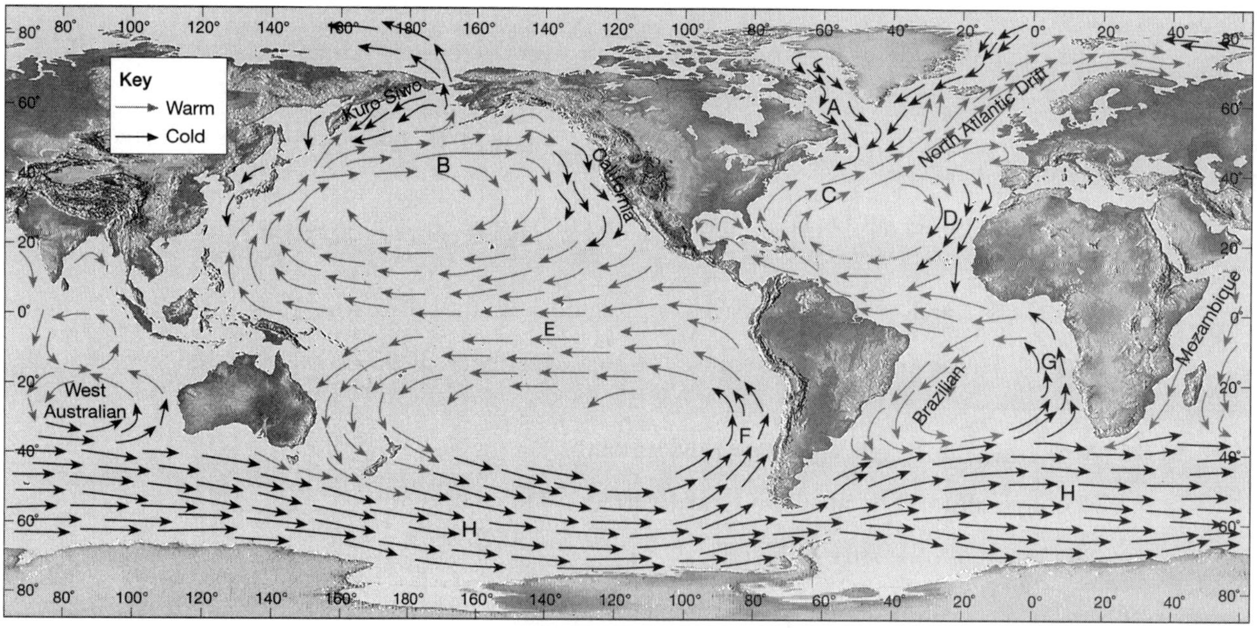

Figure 10.1

3. Describe the effect cold ocean currents have on climates.

4. In addition to the gravitational influence of the sun and moon, what are two other factors that influence the tide?

   1)

   2)

5. Using Figure 10.2, select the letter of the diagram that illustrates the relation of the sun, moon, and Earth for each of the following types of tides.

   a) Neap tide: ____ b) Spring tide: ____

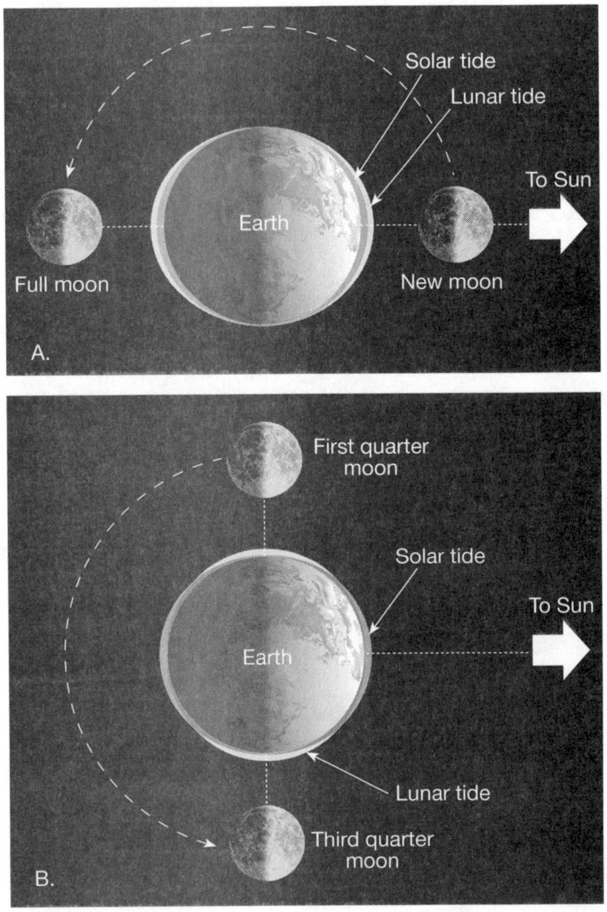

Figure 10.2

6. Using Figure 10.3, write the proper term for the part of the wave illustrated by each of the following letters.

   a) Letter A: ____________ c) Letter C: ____________

   b) Letter B: ____________ d) Letter D: ____________

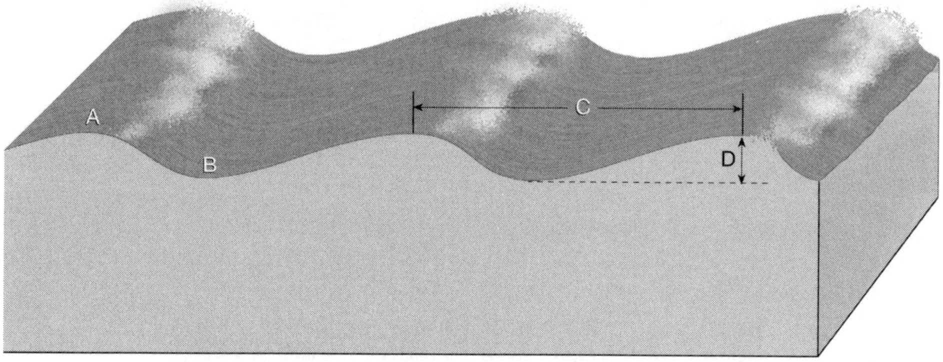

Figure 10.3

7. List the three factors that influence the height, length, and period of open water waves.
   1)
   2)
   3)

8. What is the difference between a wave of oscillation and a wave of translation?

9. What two factors are most significant in creating a dense mass of ocean water?
   1)
   2)

10. Where are the two major regions where dense water masses are created?
    1)
    2)

11. Describe the effect that wave refraction has on the energy of waves in a bay.

12. List two mechanisms which are responsible for transporting sediment along a coast.
    1)
    2)

13. Figure 10.4 illustrates the changes that take place through time along an initially irregular, relatively stable coastline. Using Figure 10.4, select the letter that illustrates each of the following features.

a) Baymouth bar: ____

b) Spit: ____

c) Sea stack: ____

d) Estuary: ____

e) Wave-cut cliff: ____

f) Tombolo: ____

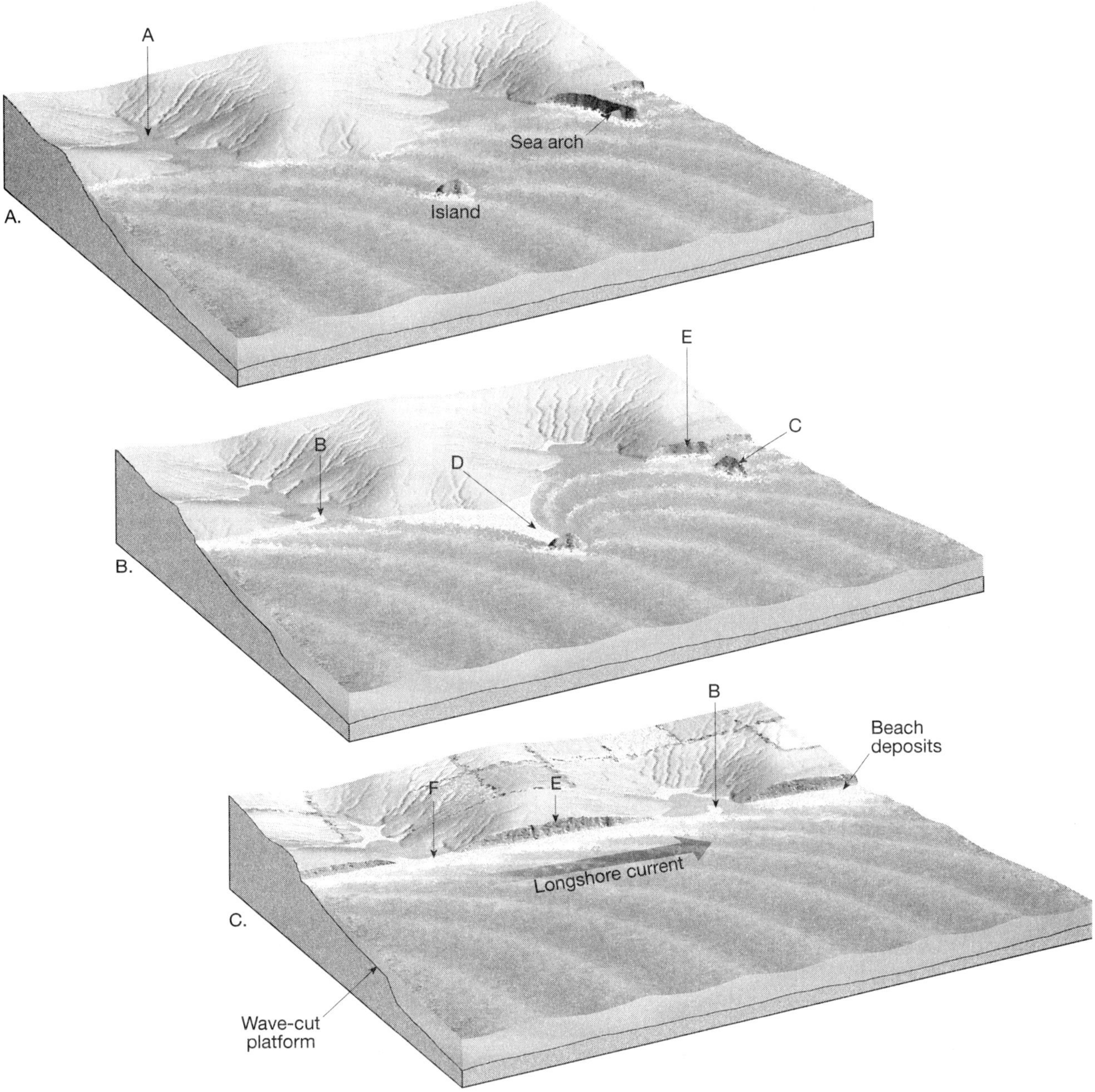

Figure 10.4

14. What are two ways that barrier islands may originate?
    1)

    2)

15. What are three factors that influence the type and degree of shoreline erosion?
    1)

    2)

    3)

16. Briefly describe the conditions that could result in the formation of the following types of coasts. Also list one feature you would expect to find associated with each type.

    a) Emergent coast:

    Feature: ____________

    b) Submergent coast:

    Feature: ____________

Practice Test

Multiple choice. Choose the best answer for the following multiple choice questions.

1. The movement of particles of sediment in a zigzag pattern along a beach is known as ____________.

   a) longshore transportation c) turbidity current e) fetch
   b) beach drift d) beach abrasion

2. A barrier built at a right angle to the beach to trap sand that is moving parallel to the shore is known as a(n) ____________.

   a) seawall c) pier e) headland
   b) groin d) stack

3. A ridge of sand that connects an island to the mainland or to another island is called a ____________.

   a) baymouth bar c) sea stack e) spit
   b) sea arch d) tombolo

4. Where the atmosphere and ocean are in contact, energy is passed from the moving air to the water through ____________.

   a) friction c) oscillation e) translation
   b) radiation d) collision

5. Waves in the open sea are called waves of ___________.
   a) translation c) modulation e) oscillation
   b) motion d) combination

6. Which one of the following is NOT a west coast desert influenced by cold offshore waters?
   a) Namib b) Gobi c) Atacama

7. Exposed wave-cut cliffs and platforms are frequently found along ___________ coasts.
   a) submergent b) emergent

8. The densest water in all of the oceans is formed in ___________ waters.
   a) Equatorial c) North Pacific e) Indian Ocean
   b) Antarctic d) North Atlantic

9. The energy and motion of most waves is derived from ___________.
   a) currents c) gravity e) earthquakes
   b) tides d) wind

10. Other than wind, the most significant factor that influences the movement of ocean water is the ___________.
    a) translation force c) force of gravity e) ocean slope
    b) Coriolis effect d) density effect

11. Which one of the following features occurs more frequently along the Atlantic and Gulf Coasts than along the Pacific Coast?
    a) wave-cut cliffs b) barrier islands c) sea stacks d) wave-cut platforms

12. Tidal currents that advance into the coastal zone as the tide rises are called ___________ currents.
    a) density c) flow e) ebb
    b) turbidity d) longshore

13. Equatorial currents derive their energy principally from the ___________ winds.
    a) easterly c) monsoon e) trade
    b) westerly d) chinook

14. In the open sea, the movement of water particles in a wave becomes negligible at a depth equal to about ___________ the distance from crest to trough.
    a) one-fourth c) one-half e) three-fourths
    b) one-third d) two-thirds

15. Deep-ocean circulation is referred to as ___________ circulation.
    a) density c) ocean-floor e) abyssal
    b) subsurface d) thermohaline

16. When a wave breaks it becomes a wave of ___________.
    a) translation c) modulation e) oscillation
    b) motion d) combination

17. The areas affected by alternating tidal currents are known as tidal ___________.
    a) swamps c) flats e) estuaries
    b) marshes d) platforms

18. Daily changes in the elevation of the ocean surface are called ____________.
   a) currents
   b) waves
   c) density currents
   d) tides
   e) estuaries

19. The large central area of the North Atlantic which has no well-defined currents is known as the ____________ Sea.
   a) Red
   b) Baltic
   c) Sargasso
   d) Central Atlantic
   e) Central

20. Highly irregular coasts with numerous estuaries frequently characterize a(n) ____________ coast.
   a) submergent
   b) emergent

21. The horizontal distance separating wave crests is known as the ____________.
   a) wavelength
   b) fetch
   c) wave height
   d) wave period
   e) oscillation

22. The largest daily tidal ranges are associated with ____________ tides.
   a) spring
   b) ebb
   c) neap
   d) flow
   e) lunar

23. The time interval between the passage of successive wave crests is called the ____________.
   a) wave height
   b) wavelength
   c) wave period
   d) fetch
   e) oscillation

24. A wave begins to "feel bottom" at a water depth equal to about ____________ its wavelength.
   a) one-fourth
   b) one-third
   c) one-half
   d) two-thirds
   e) three-fourths

25. Massive barriers built to prevent waves from reaching the shore are called ____________.
   a) groins
   b) seawalls
   c) stacks
   d) barrier islands
   e) piers

True/false. For the following true/false questions, if a statement is not completely true, mark it false. For each false statement, change the ***italicized*** *word to correct the statement.*

1. ___ Turbulent water currents within the surf zone that flow parallel to the shore are called *longshore* currents.

2. ___ Evaporation and/or freezing tend to make the water *less* salty.

3. ___ *Reach* is the distance that wind has traveled across the open water.

4. ___ The two most significant factors in creating a dense mass of water are temperature and *salinity.*

5. ___ In the open sea, it is the wave *form* that moves forward, not the water itself.

6. ___ In the Southern Hemisphere, ocean currents form *counterclockwise* gyres.

7. ___ Groins have proven to be *unsatisfactory* solutions to shoreline erosion problems.

8. ___ The tide-generating potential of the sun is slightly less than *half* that of the moon.

9. ___ Two caves on opposite sides of a headland can unite to form a sea *cliff.*

10. ___ *Cold* ocean currents often have a dramatic impact on the climate of tropical deserts along the west coasts of continents.

11. ___ Ocean currents transfer *excess* heat from the tropics to the heat-deficient polar regions.

12. ___ Wind speeds less than *three* kilometers per hour generally do not produce stable, progressive waves.

13. ___ In the Northern Hemisphere, the Coriolis effect causes ocean currents to be deflected to the *left.*

14. ___ Turbulent water created by breaking waves is called *drift.*

15. ___ The lowest daily tidal ranges are associated with *neap* tides.

Written questions

1. Describe coastal upwelling and the affect it has on fish populations.

2. Briefly describe thermohaline circulation.

3. What is the cause of ocean tides?

4. Describe the motion of a surface water particle in a wave of oscillation and in a wave of translation.

11

Heating the Atmosphere

Heating the Atmosphere introduces the subject of meteorology by presenting its definition, noting the differences between weather and climate, and listing the elements of weather. Included is a discussion of the chemical composition of the atmosphere and the depletion of ozone. The structure and extent of the atmosphere are also examined.

Atmospheric heating begins by examining Earth's motions. The variables that control the quantityof solar radiation intercepted by a particular place are discussed. This is followed by a detailed investigation of the seasons. Radiant energy and common mechanisms of heat transfer are introduced, which leads to a discussion of atmospheric heating by solar and terrestrial radiation. The factors that cause variations in temperature, such a, differential heating of land and water, altitude, and geographic position, are explained. The chapter concludes with a short discussion of the global distribution of surface temperatures.

Learning Objectives

After reading, studying, and discussing this chapter, you should be able to:

- Describe the science of meteorology.
- Explain the difference between weather and climate.
- List the most important elements of weather and climate.
- List the major components of clean, dry air.
- Describe the extent and structure of the atmosphere.
- Describe how the atmosphere is heated.
- List the factors that cause temperature to vary from place to place.
- Explain the causes of the seasons.
- Describe the general distribution of global surface temperatures.

Chapter Review

- *Weather* is the state of the atmosphere at a particular place for a short period of time. *Climate*, on the other hand, is a generalization of the weather conditions of a place over a long period of time.

- The most important *elements*, those quantities or properties that are measured regularly, of weather and climate are 1) air *temperature*, 2) *humidity*, 3) type and amount of *cloudiness*, 4) type and amount of *precipitation*, 5) air *pressure*, and 6) the speed and direction of the *wind*.

- If water vapor, dust, and other variable components of the atmosphere were removed, clean, dry air would be composed almost entirely of *nitrogen* (N), about 78% of the atmosphere by volume, and *oxygen* (O_2), about 21%. *Carbon dioxide* (CO_2), although present only in minute is important because it has the ability to absorb heat radiated by Earth and thus helps keep the atmosphere warm. Among the variable components of air, *water vapor* is very important because it is the source of all clouds and precipitation and, like carbon dioxide, it is also a heat absorber.

• *Ozone* (O_3), the triatomic form of oxygen, is concentrated in the 10- to 50-kilometer altitude range of the atmosphere, and is important to life because of its ability to absorb potentially harmful ultraviolet radiation from the sun.

• Because the atmosphere gradually thins with increasing altitude, it has no sharp upper boundary but simply blends into outer space. Based on temperature, the atmosphere is divided vertically into four layers. The *troposphere* is the lowermost layer. In the troposphere, temperature usually decreases with increasing altitude. This *environmental lapse rate* is variable, but averages about 6.5°C per kilometer (3.5°F per 1000 feet). Essentially all important weather phenomena occur in the troposphere. Above the troposphere is the *stratosphere*, which exhibits warming because of absorption of ultraviolet radiation by ozone. In the *mesosphere*, temperatures again decrease. Upward from the mesosphere is the *thermosphere*, a layer with only a minute fraction of the atmosphere's mass and no well-defined upper limit.

• The two principal motions of Earth are 1) *rotation*, the spinning of Earth about its axis, which produces the daily cycle of daylight and darkness, and 2) *revolution*, the movement of Earth in its orbit around the sun.

• Several factors act together to cause the seasons. Earth's axis is inclined 23 1/2° from the perpendicular to the plane of its orbit around the sun and remains pointed in the same direction (toward the North Star) as Earth journeys around the sun. As a consequence, Earth's orientation to the sun continually changes. The yearly fluctuations in the angle of the sun and length of daylight brought about by Earth's changing orientation to the sun cause seasons.

• The three mechanisms of heat transfer are 1) *conduction*, the transfer of heat through matter by molecular activity, 2) *convection*, the transfer of heat by the movement of a mass or substance from one place to another, and 3) *radiation*, the transfer of heat by electromagnetic waves.

• *Electromagnetic radiation* is energy emitted in the form of rays, or waves, called electromagnetic waves. All radiation is capable of transmitting energy through the vacuum of space. One of the most important differences between electromagnetic waves are their *wavelengths*, which range from very long *radio waves* to very short *gamma rays*. *Visible light* is the only portion of the electromagnetic spectrum we can see. Some of the basic laws that govern radiation as it heats the atmosphere are 1) all objects emit radiant energy, 2) hotter objects radiate more total energy than do colder objects, 3) the hotter the radiating body, the shorter the wavelengths of maximum radiation, and 4) objects that are good absorbers of radiation are good emitters as well.

• The general drop in temperature with increasing altitude in the troposphere supports the fact that *the atmosphere is heated from the ground up*. Approximately 50% of the solar energy, primarily in the form of the shorter wavelengths, that strikes the top of the atmosphere is ultimately absorbed at Earth's surface. Earth releases the absorbed radiation in the form of long-wave radiation. The atmospheric absorption of this long-wave *terrestrial radiation*, primarily by water vapor and carbon dioxide, is responsible for heating the atmosphere.

• Carbon dioxide, an important heat absorber in the atmosphere, is one of several gases that influence *global warming*. Some consequences of global warming could be 1) shifts in temperature and rainfall patterns, 2) a gradual rise in sea level, 3) changing storm tracks and both the higher frequency and greater intensity of hurricanes, and 4) an increase in the frequency and intensity of heat waves and droughts.

- The factors that cause temperature to vary from place to place, also called the *controls of temperature*, are 1) differences in the *receipt of solar radiation*—the greatest single cause, 2) the unequal heating and cooling of *land and water*, in which land heats more rapidly and to higher temperatures than water and cools more rapidly and to lower temperatures than water, 3) *altitude*, 4) *geographic position*, 5) *cloud cover and albedo*, and 6) *ocean currents*.
- Temperature distribution is shown on a map by using *isotherms*, which are lines that connect equal temperatures.

Chapter Outline

I. Weather and climate
  A. Weather
    1. Weather is over a short period of time
    2. Constantly changing
  B. Climate
    1. Climate is over a long period of time
    2. Generalized, composite of weather
  C. Elements of
    1. Properties that are measured regularly
    2. Most important elements
      a. Temperature
      b. Humidity
      c. Cloudiness
      d. Precipitation
      e. Air pressure
      f. Wind speed and direction
II. Composition of the atmosphere
  A. Air is a mixture of discrete gases
  B. Composition of clean, dry air
    1. Nitrogen (N)—78%
    2. Oxygen (O_2)—21%
    3. Argon and other gases
    4. Carbon dioxide (CO_2)—0.036% (absorbs heat energy)
  C. Variable components
    1. Water vapor
      a. Up to about 4% of air volume
      b. Forms clouds and precipitation
      c. Absorbs heat energy from Earth
    2. Dust
      a. Includes pollen and spores
      b. Water vapor condenses on
      c. Reflects sunlight
      d. Helps color sunrise and sunset
  D. Ozone
    1. Three atom oxygen (O_3)
    2. Distribution not uniform
    3. Concentrated between 10 to 50 km
    4. Absorbs harmful UV radiation
    5. Human activity is depleting ozone by adding chlorofluorocarbons
III. Structure of the atmosphere
  A. Layered structure based on temperature
  B. Layers
    1. Troposphere
      a. Bottom layer
      b. Temperature decrease with altitude is called the environmental lapse rate
        1. 6.5°C per kilometer (average)
        2. 3.5°F per 1000 feet (average)
      c. Thickness varies—average height about 12 km
      d. Outer boundary is called the tropopause
    2. Stratosphere
      a. About 12 km to 50 km
      b. Temperature increases at top
      c. Outer boundary is called the stratopause
    3. Mesosphere
      a. About 50 km to 80 km
      b. Temperature decreases
      c. Outer boundary—mesopause
    4. Thermosphere
      a. No well-defined upper limit
      b. Fraction of atmosphere's mass
      c. Gases moving at high speeds
IV. Earth-sun relations
  A. Earth motions
    1. Rotates on its axis
    2. Revolves around the sun

- B. Seasons
  - 1. Result of
    - a. Changing sun angle
    - b. Changing length of daylight
  - 2. Caused by earth's changing orientation to sun
    - a. Axis is inclined 23 1/2°
    - b. Axis always in same direction
  - 3. Special days (Northern Hemisphere)
    - a. Summer solstice
      - 1. June 21-22
      - 2. Location of sun's vertical rays
        - a. Tropic of Cancer
        - b. 23 1/2°N latitude
    - b. Winter solstice
      - 1. December 21-22
      - 2. Location of sun's vertical rays
        - a. Tropic of Capricorn
        - b. 23 1/2°S latitude
    - c. Autumnal equinox
      - 1. September 22-23
      - 2. Location of sun's vertical rays
        - a. Equator
        - b. 0° latitude
    - d. Spring equinox
      - 1. March 21-22
      - 2. Location of sun's vertical rays
        - a. Equator
        - b. 0° latitude

V. Atmospheric heating
- A. Mechanisms of heat transfer
  - 1. Conduction—through molecular activity
  - 2. Convection
    - a. Movement of a mass
    - b. Usually vertical motions
    - c. Horizontal motions called advection
- B. Radiation (electromagnetic radiation)
  - 1. 186,000 miles per second in a vacuum
  - 2. Different wavelengths
    - a. Gamma (very short)
    - b. X-rays
    - c. Ultraviolet (UV)
    - d. Visible
    - e. Infrared
    - f. Microwaves
    - g. Radio (longest)
  - 3. Governed by basic laws
- C. Incoming solar radiation
  - 1. Atmosphere largely transparent to
  - 2. Atmospheric effects
    - a. Scattering
    - b. Reflection albedo (percent reflected)
    - c. Absorption
  - 3. Most visible radiation reaches surface
  - 4. 50% absorbed at Earth's surface
- D. Radiation from Earth's surface
  - 1. Earth reradiates at longer wavelengths
  - 2. Longer terrestrial radiation absorbed by
    - a. Carbon dioxide
    - b. Water vapor
  - 3. Lower atmosphere heated from surface
- E. Global warming
  - 1. Caused by increased atmospheric CO_2
  - 2. Possible future increase of 1.5 – 3°C
  - 3. Aided by other trace gases

VI. Temperature measurement
- A. Daily maximum and minimum
- B. Other measurements
  - 1. Daily mean temperature
  - 2. Daily range
  - 3. Monthly mean
  - 4. Annual mean
  - 5. Annual temperature range

VII. Controls of temperature
- A. Cause temperature to vary
- B. Receipt of solar radiation—most important control
- C. Other important controls
  - 1. Differential heating of land and water
    - a. Land heats more rapidly than water
    - b. Land gets hotter than water
    - c. Land cools faster than water
    - d. Land gets cooler than water
  - 2. Altitude
  - 3. Geographic position
  - 4. Cloud cover and albedo
  - 5. Ocean currents

VIII. World distribution of temperature
- A. Temperature maps
  - 1. Isotherm—line connecting places of equal temperature

2. Temperatures adjusted to sea level
3. January and July used for analysis —represent temperature extremes

B. Global temperature patterns
1. Decrease poleward from tropics
2. Latitudinal shift with seasons
3. Warmest and coldest over land
4. In Southern Hemisphere
   a. Isotherms straighter
   b. Isotherms more stable
5. Isotherms show ocean currents
6. Annual temperature range
   a. Small near equator
   b. Increases with increase in latitude
   c. Greater over continental locations

Vocabulary Review

Choosing from the list of key terms, furnish the most appropriate response for the following statements.

1. The bottom layer of the atmosphere, where temperature decreases with an increase in altitude is the ____________.

2. ____________ is a word used to denote the state of the atmosphere at a particular place for a short period of time.

3. ____________ is the spinning of Earth about its axis.

4. The outermost layer of the atmosphere is the ____________.

5. ____________ might be best described as an aggregate or composite of weather.

6. A line on a map connecting points of equal temperature is called a(n) ____________.

7. The ____________ is located at 23 1/2 degrees north latitude.

8. A quantity or property of weather and climate that is measured regularly is termed a(n) ____________.

9. ____________ is the transfer of heat by the movement of a mass or substance from one place to another, usually vertically.

10. The fact that Earth's axis is not perpendicular to the plane of its orbit is referred to as the ____________.

11. The time when the vertical rays of the sun strike the equator is known as the ____________.

12. The temperature decrease with increasing altitude in the troposphere is called the ____________.

13. The ____________ is located at 23 1/2 degrees south latitude.

14. The great circle that separates daylight from darkness is the ____________.

15. In the atmospheric layer known as the ____________, temperatures decrease with height until approximately 80 kilometers above the surface.

16. The transfer of energy through space by electromagnetic waves is known as ____________.

17. The movement of Earth in its orbit around the sun is referred to as ____________.

Key Terms

Page numbers shown in ( ) refer to the textbook page where the term first appears.

air (p. 272)
albedo (p. 286)
circle of illumination (p. 278)
climate (p. 272)
conduction (p. 282)
convection (p. 283)
element (of weather and climate) (p. 272)
environmental lapse rate (p. 276)
equinox (spring or autumnal) (p. 280)
greenhouse effect (p. 287)
inclination of the axis (p. 279)
infrared (p. 284)
isotherm (p. 291)
mesosphere (p. 277)
radiation (electromagnetic radiation) (p. 283)
revolution (p. 278)
rotation (p. 278)
solstice (summer or winter) (p. 280)
stratosphere (p. 277)
thermosphere (p. 277)
Tropic of Cancer (p. 280)
Tropic of Capricorn (p. 280)
troposphere (p. 276)
ultraviolet (UV) (p. 284)
visible light (p. 284)
weather (p. 272)

18. ____________ is the transfer of heat through matter by molecular activity.

19. The atmospheric layer immediately above the tropopause is the ____________.

20. Although we cannot see it, we detect ____________ radiation as heat.

21. The time when the vertical rays of the sun are striking either the Tropic of Cancer or the Tropic of Capricorn is known as the ____________.

22. Intense exposure to ____________ radiation is responsible for sunburn.

23. The percentage of the total radiation that is reflected by a surface is called its ____________.

24. The ____________ refers to the mechanism responsible for heating the atmosphere.

Comprehensive Review

1. Distinguish between the two terms *weather* and *climate.*

2. Using Figure 11.1, select the letter that indicates each of the following parts of Earth's atmosphere.

   a) Mesosphere: ____
   d) Troposphere: ____

   b) Tropopause: ____
   e) Stratopause: ____

   c) Stratosphere: ____
   f) Thermosphere: ____

3. List and describe the two principal motions of Earth.

   1)

   2)

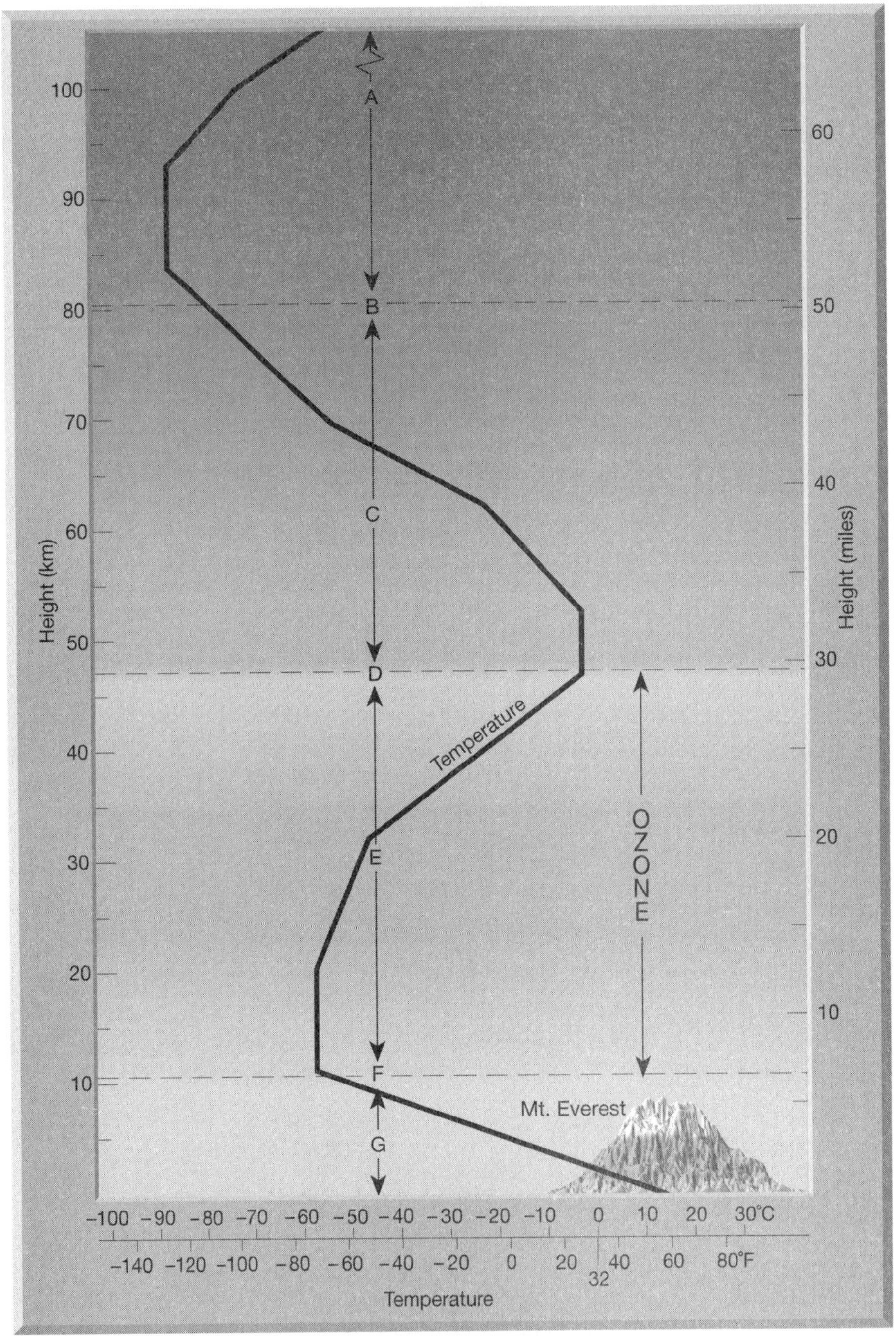
Height (km)
100
90
80
70
60
50
40
30
20
10
Height (miles)
60
50
40
30
20
10
A
B
C
D
E
F
G
Temperature
OZONE
Mt. Everest
-100 -90 -80 -70 -60 -50 -40 -30 -20 -10 0 10 20 30°C
-140 -120 -100 -80 -60 -40 -20 0 20 40 60 80°F
32
Temperature

Figure 11.1

4. List the six most important elements of weather and climate.

   1) 4)

   2) 5)

   3) 6)

5. List two of the meteorological roles of atmospheric dust.

   1)

   2)

6. What are the two ways that the seasonal variation in the altitude of the sun affect the amount of solar energy received at Earth's surface?

   1)

   2)

7. Using Figure 11.2, list the Northern Hemisphere season, date, and latitude of the vertical sun for each of the four lettered positions.

   a) Position A:

   b) Position B:

   c) Position C:

   d) Position D:

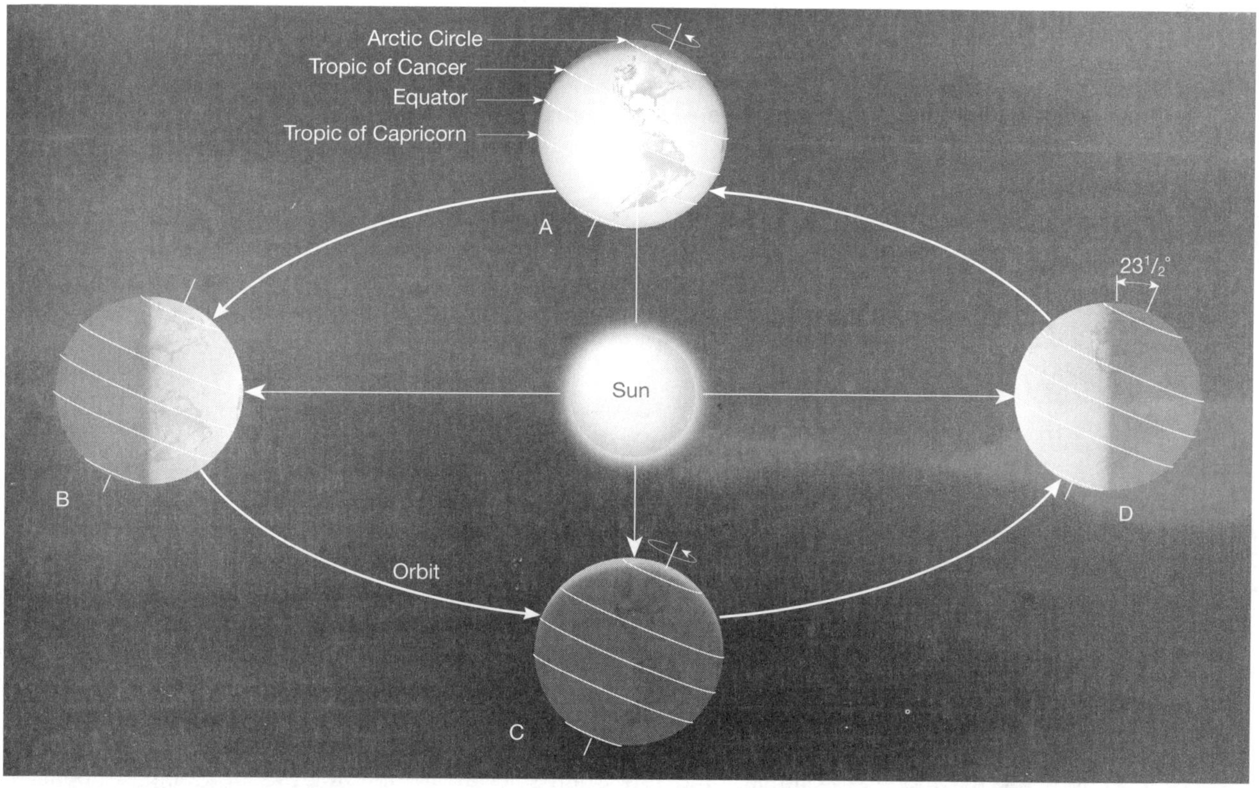

Figure 11.2

8. What are the two most abundant gases in clean, dry air? What are their percentages?

9. How does ozone differ from oxygen? What function does atmospheric ozone serve, especially for those of us on Earth?

10. In an average situation, approximately what percentage of incoming solar radiation is
    a) Absorbed by the atmosphere and clouds: ____%

    b) Absorbed by Earth's surface: ____%

    c) Reflected by clouds: ____%

11. Beginning with the fact that more incoming solar radiation is absorbed by Earth's surface than by the atmosphere, describe the mechanism responsible for atmospheric heating.

12. Which two atmospheric gases are the principal absorbers of heat energy?

13. List and briefly describe the three mechanisms of heat transfer.
    1)

    2)

    3)

14. What are three possible consequences of increasing global temperatures?
    1)

    2)

    3)

15. Briefly explain how each one of the following temperatures are determined:
    a) Daily mean temperature:

    b) Annual mean temperature:

    c) Annual temperature range:

16. What is the affect of each of the following on temperature?
    a) Differential heating of land and water:

    b) Altitude:

    c) Geographic position:

17. Why are the months of January and July often selected for the analysis of global temperature patterns?

Practice Test

Multiple choice. Choose the best answer for the following multiple choice questions.

1. Which one of the following is NOT generally considered a major element of weather and climate?
   a) ocean currents c) air temperature e) wind speed
   b) humidity d) air pressure

2. A form of oxygen that combines three oxygen atoms into each molecule is called ____________.
   a) trioxygen c) oxulene e) azide
   b) oxyurid d) ozone

3. Oxygen and ozone are efficient absorbers of incoming ____________ radiation.
   a) ultraviolet c) radio e) gamma
   b) infrared d) visible

4. The generalization of atmospheric conditions over a long period of time is referred to as ____________.
   a) meteorology c) global synthesis e) element analysis
   b) climate d) weather

5. On June 21, the vertical rays of the sun strike a line of latitude known as the ____________.
   a) equator c) Tropic of Capricorn e) Antarctic Circle
   b) Arctic Circle d) Tropic of Cancer

6. The red and orange colors of sunset and sunrise are the result of ____________ in the atmosphere.
   a) moisture c) insects e) argon
   b) dust d) pollen

7. The bottom layer of the atmosphere in which we live is called the ____________.
   a) mesosphere c) troposphere e) stratosphere
   b) thermosphere d) hemisphere

8. Which one of the following forms of radiation has the longest wavelength?
   a) radio waves c) blue light e) infrared
   b) ultraviolet d) yellow light

9. Ozone in the atmosphere is concentrated in a layer called the ____________.
   a) troposphere c) mesosphere e) hemisphere
   b) stratosphere d) thermosphere

10. Which one of the following is NOT a mechanism of heat transfer?
    a) conduction b) condensation c) radiation d) convection

11. The temperature decrease with altitude in the troposphere is called the ____________ lapse rate.
    a) environmental c) ascending e) incremental
    b) altitude d) atmospheric

12. On December 21, the vertical rays of the sun strike a line of latitude known as the ____________.
    a) equator c) Tropic of Capricorn e) Antarctic Circle
    b) Arctic Circle d) Tropic of Cancer

13. Which one of the following gases has the ability to absorb heat energy radiated by Earth?
   a) oxygen c) ozone e) carbon dioxide
   b) helium d) nitrogen

14. The chemicals contributing to the depletion of atmospheric ozone are referred to as ___________.
   a) CCCs c) FCDs e) CFDs
   b) CFCs d) PCBs

15. Approximately what percentage of the solar energy that strikes the top of the atmosphere reaches Earth's surface?
   a) 30% c) 50% e) 70%
   b) 40% d) 60%

16. Which one of the following is NOT an important control of temperature?
   a) heating of land and water c) wind e) ocean currents
   b) altitude d) geographic position

17. The two principal atmospheric absorbers of terrestrial radiation are carbon dioxide and ___________.
   a) oxygen c) nitrogen e) helium
   b) water vapor d) ozone

18. When water changes from one state to another, ___________ heat is stored or released.
   a) latent c) fusion e) elemental
   b) vapor d) reduced

19. Which one of the following yearly fluctuations is NOT responsible for the seasons?
   a) length of daylight b) sun angle c) Earth-sun distance

20. The value of the normal lapse rate is ___________C degrees per kilometer.
   a) 3.5 c) 5.5 e) 7.5
   b) 4.5 d) 6.5

21. Ninety percent of the atmosphere lies below ___________ miles.
   a) 5 c) 15 e) 25
   b) 10 d) 20

22. In the Southern Hemisphere, the greatest number of daylight hours occurs on ___________.
   a) June 21 c) March 21 e) September 21
   b) December 21 d) April 21

23. On the equinoxes, the vertical rays of the sun strike a line of latitude known as the ___________.
   a) equator c) Tropic of Capricorn e) Antarctic Circle
   b) Arctic Circle d) Tropic of Cancer

24. The amount of energy received at Earth's surface is controlled by the ___________ of the sun.
   a) temperature c) distance e) rotation
   b) altitude d) color

25. Temperature distribution is shown on a map using lines called __________ that connect places of equal temperature.
    a) isotemps b) equagrads c) isobars d) isotherms

26. Which form of radiation do we detect as heat?
    a) radio waves c) red light e) infrared
    b) ultraviolet d) yellow light

27. Combustion of fossil fuels adds vast quantities of the gas __________ to the atmosphere.
    a) oxygen c) ozone e) carbon dioxide
    b) water vapor d) nitrogen

28. The annual temperature range will be greatest for a place located __________.
    a) at the pole c) in the interior of a continent e) near the coast
    b) on the equator d) at high altitude

29. Which one of the following surfaces has the highest albedo?
    a) concrete c) soil e) forest
    b) snow d) water

30. The most abundant gas in clean, dry air is __________.
    a) oxygen c) nitrogen e) ozone
    b) argon d) carbon dioxide

True/false. For the following true/false questions, if a statement is not completely true, mark it false. For each false statement, change the ***italicized*** *word to correct the statement.*

1. ___ In the Northern Hemisphere, December 21 is referred to as the *winter* solstice.
2. ___ *Climate* is a word used to denote the state of the atmosphere at a particular place over a short period of time.
3. ___ The *higher* the solar angle or altitude of the sun, the more spread out and less intense is the solar radiation that reaches Earth's surface.
4. ___ On the *equinoxes*, all places on Earth receive 12 hours of daylight.
5. ___ Atmospheric ozone is an absorber of potentially harmful *ultraviolet* radiation from the sun.
6. ___ The transfer of heat through matter by molecular activity is called *radiation.*
7. ___ An international agreement known as the *Montreal Protocol* established a schedule for eliminating CFCs from general use.
8. ___ The most important difference among electromagnetic waves is their *wavelength.*
9. ___ If Earth's axis were not *inclined*, we would have no seasons.
10. ___ The atmosphere is divided into four layers on the basis of *composition.*

11. ___ The *higher* the solar angle or altitude of the sun, the less distance solar radiation must penetrate through the atmosphere to Earth's surface.

12. ___ Essentially all important weather occurs in the *stratosphere.*

13. ___ The line separating the dark half of Earth from the lighted half is called circle of *daylight.*

14. ___ In the Southern Hemisphere, June 21 is referred to as the *summer* solstice.

15. ___ There is a *sharp* boundary between the atmosphere and outer space.

16. ___ The hotter the radiating body, the *shorter* the wavelength of maximum radiation.

17. ___ Earth's two principal motions are rotation and *revolution.*

18. ___ The *visible* wavelengths of radiation from the sun reach Earth's surface.

19. ___ Heating an object causes its atoms to move *slowly.*

20. ___ The percentage of the total radiation that is reflected by an object is called its *albedo.*

21. ___ The principal source for atmospheric heat is *terrestrial* radiation.

22. ___ Land heats *more* rapidly than water.

23. ___ Compared the Northern Hemisphere, the Southern Hemisphere has a *greater* annual temperature variation.

Written questions

1. What is the relation between CFCs and the ozone problem?

2. What causes the amount of solar energy reaching places on Earth's surface to vary with the seasons?

3. Briefly describe how Earth's atmosphere is heated.

12

Clouds and Precipitation

Clouds and Precipitation begins with an examination of the processes and energy requirements involved in the changes of state of water, including the ability of water to store and release latent heat. After presenting the various methods used to express humidity, the relations between temperature, water vapor content, relative humidity, and dew-point temperature are explored. Also investigated is the adiabatic process and condensation aloft. Following a detailed explanation of the conditions that influence the stability of air, the three processes that lift air are described. Included in the chapter is the classification of clouds as well as the formation and types of fog. The chapter concludes with a discussion of the processes involved in the formation of precipitation and the various forms of precipitation.

Learning Objectives

After reading, studying, and discussing this chapter, you should be able to:

- List the processes that cause water to change from one of state of matter to another.
- Explain saturation, vapor pressure, specific humidity, relative humidity, and dew point.
- Describe how relative humidity is determined.
- Explain the basic cloud-forming process.
- Describe stable and unstable air.
- List the three processes that initiate the vertical movement of air.
- Discuss the conditions necessary for condensation.
- List the criteria used to classify clouds.
- Describe the formation of fog.
- Discuss the formation and forms of precipitation.

Chapter Review

- *Water vapor*, an odorless, colorless gas, changes from one state of matter (solid, liquid, or gas) to another at the temperatures and pressures experienced near Earth's surface. The processes involved in *changing the state of matter* of water are *evaporation*, *condensation*, *melting*, *freezing*, *sublimation*, and *deposition*.
- *Humidity* is the general term used to describe the amount of water vapor in the air. *Relative humidity*, the *ratio* (expressed as a percent) of the air's water vapor *content* to its water vapor *capacity* at a given temperature, is the most familiar term used to describe humidity. The water vapor capacity of air is temperature dependent, with warm air having a much greater capacity than cold air.

• *Relative humidity can be changed in two ways.* One is by *adding* or *subtracting water vapor.* The second is by *changing the air's temperature.* When air is cooled, its relative humidity increases. Air is said to be saturated when it contains the maximum quantity of water vapor that it can hold at any given temperature and pressure. *Dew point* is the temperature to which air would have to be cooled in order to reach saturation.

• The cooling of air as it rises and expands due to successively lower pressure is the basic cloud-forming process. Temperature changes in air brought about by compressing or expanding the air are called *adiabatic temperature changes.* Unsaturated air warms by compression and cools by expansion at the rather constant rate of 10°C per 1000 meters of altitude change, a figure called the *dry adiabatic rate.* If air rises high enough it will cool sufficiently to cause condensation and form a cloud. From this point on, air that continues to rise will cool at the *wet adiabatic rate* which varies from 5°C to 9°C per 1000 meters of ascent. The difference in the wet and dry adiabatic rates is caused by the condensing water vapor releasing *latent heat,* thereby reducing the rate at which the air cools.

• The *stability of air* is determined by examining the temperature of the atmosphere at various altitudes. Air is said to be *unstable* when the *environmental lapse rate* (the rate of temperature decrease with increasing altitude in the troposphere) is greater than the *dry adiabatic rate.* Stated differently, a column of air is unstable when the air near the bottom is significantly warmer (less dense) than the air aloft.

• Three mechanisms that can initiate the vertical movement of air are 1) *orographic lifting,* which occurs when elevated terrains, such as mountains, act as barriers to the flow of air; 2) *frontal wedging,* when cool air acts as a barrier over which warmer, less dense air rises; and 3) *convergence,* which happens when air flows together and a general upward movement of air occurs.

• For condensation to occur, air must be saturated. Saturation takes place either when air is cooled to its dew point, which most commonly happens, or when water vapor is added to the air. There must also be a surface on which the water vapor may condense. In cloud and fog formation, tiny particles called *condensation nuclei* serve this purpose.

• *Clouds* are classified on the basis of their *appearance* and *height.* The three basic forms are *cirrus* (high, white, thin, wispy fibers), *cumulus* (globular, individual cloud masses), and *stratus* (sheets or layers that cover much or all of the sky). The four categories based on height are *high clouds* (bases normally above 6000 meters), middle clouds (from 2000 to 6000 meters), *low clouds* (below 2000 meters), and *clouds of vertical development.*

• *Fog* is defined as a cloud with its base at or very near the ground. Fogs form when air is cooled below its dew point or when enough water vapor is added to the air to bring about saturation. Various types of fog include *advection fog, radiation fog, upslope fog, steam fog,* and *frontal* (or *precipitation*), fog.

• For *precipitation* to form, millions of cloud droplets must somehow join together into large drops. Two mechanisms for the formation of precipitation have been proposed. One, in clouds where the temperatures are below freezing, ice crystals form and fall as snowflakes. At lower altitudes the snowflakes melt and become raindrops before they reach the ground. Two, large droplets form in warm clouds that contain large *hygroscopic* ("water seeking") *nuclei,* such as salt particles. As these big droplets descend, they collide and join with smaller water droplets. After many collisions the droplets are large enough to fall to the ground as rain.

• The forms of precipitation include *rain, snow, sleet, hail,* and *rime.*

Chapter Outline

- I. Changes of state of water
  - A. Three states of matter
    - 1. Solid
    - 2. Liquid
    - 3. Gas
  - B. To change state, heat must be
    - 1. Absorbed, or
    - 2. Released
  - C. Heat energy
    - 1. Measured in calories—heat necessary to raise the temperature of 1 gram of water 1°C
    - 2. Latent heat
      - a. Stored or hidden heat
      - b. Not derived from temperature change
      - c. Important in atmospheric processes
  - D. Processes
    - 1. Evaporation
      - a. Liquid changed to gas
      - b. 600 calories per gram of water added (latent heat of vaporization)
    - 2. Condensation
      - a. Water vapor (gas) changed to liquid
      - b. Heat energy released
    - 3. Melting
      - a. Solid changed to liquid
      - b. 80 calories per gram of water added (latent heat of melting)
    - 4. Freezing
      - a. Liquid changed to solid
      - b. Heat released (latent heat of fusion)
    - 5. Sublimation
      - a. Solid changed directly to a gas (e.g., ice cubes shrinking in a freezer)
      - b. 680 calories per gram of water added
    - 6. Deposition
      - a. Water vapor (gas) changed to a solid (e.g., frost in a freezer compartment)
      - b. Heat released
- II. Humidity
  - A. Amount of water vapor in air
    - 1. Water vapor adds pressure to air (called vapor pressure)
    - 2. Saturated air is filled to capacity
    - 3. Capacity is temperature dependent—warm air has much greater capacity
  - B. Measurements of humidity
    - 1. Specific humidity
      - a. Quantity of water vapor per given mass of air
      - b. Units: grams per kilogram
    - 2. Relative humidity
      - a. Ratio of
        - 1. Actual water vapor content, to
        - 2. Potential water vapor capacity,
        - 3. At a given temperature
      - b. Expressed as a percent
      - c. Saturated air
        - 1. Content equals capacity
        - 2. 100% relative humidity
      - d. Can be changed in two ways
        - 1. Add or subtract moisture to air
          - a. Adding moisture raises the relative humidity
          - b. Removing moisture lowers the relative humidity
        - 2. Changing air temperature
          - a. Lowering temperature raises the relative humidity
          - b. Raising temperature lowers the relative humidity
      - e. Dew point
        - 1. Temperature where
          - a. Air is saturated, and
          - b. Relative humidity is 100%
        - 2. Cooling air below the dew point causes condensation
          - a. e.g., Cloud formation
          - b. Water vapor requires a surface to condense on
      - f. Instruments used to measure humidity
        - 1. Hygrometer measures humidity
        - 2. Psychrometer (a type of hygrometer)
          - a. Compares temperatures of
            - 1. Wet-bulb thermometer, and
            - 2. Dry-bulb thermometer
          - b. If air is saturated (100% relative humidity)
            - 1. Both thermometers read the same temperature

III. Adiabatic heating/cooling
   A. Adiabatic temperature changes occur when
      1. Air is compressed
         a. Motion of air molecules increases
         b. Air will warm
         c. Descending air is compressed due to increasing air pressure
      2. Air expands
         a. Air does work on surrounding air
         b. Air will cool
         c. Rising air will expand due to decreasing air pressure
   B. Adiabatic rates
      1. Dry adiabatic rate
         a. Unsaturated air
         b. Rising air expands and cools at 1°C per 100 meters
         c. Descending air is compressed and warms at 1°C per 100 meters
      2. Wet adiabatic rate
         a. Commences at condensation level
         b. Air has reached dew point
         c. Condensation is occurring (latent heat is being liberated)
         d. Heat released by condensing water reduces the rate of cooling
         e. Rate varies from 0.5°C to 0.9°C per 100 meters

IV. Stability of Air
   A. Two types
      1. Stable air
         a. Resists vertical displacement
            1. Cooler than surrounding air
            2. Denser than surrounding air
            3. Wants to sink
         b. No adiabatic cooling
         c. Stability occurs when the environmental lapse rate is less than the wet adiabatic rate
         d. Often results in widespread clouds with little vertical thickness
         e. Precipitation, if any, is light-moderate
      2. Unstable air
         a. Like hot air balloon
         b. Rising air
            1. Warmer than surrounding air
            2. Less dense than surrounding air
            3. Continues to rise until it reaches an altitude with the same temperature
         c. Adiabatic cooling
         d. Air is unstable when the environmental lapse rate is greater than the dry adiabatic rate
         e. Clouds are often towering
         f. Often results in heavy precipitation
   B. Determines to a large degree
      1. Type of clouds that develop
      2. Intensity of precipitation
   C. Processes that lift air
      1. Orographic lifting
         a. Elevated terrains act as barriers
         b. Result can be rainshadow desert
      2. Frontal wedging
         a. Cool air acts as a barrier to warm air
         b. Fronts—parts of middle-latitude cyclones
      3. Convergence—air flowing together and rising

V. Condensation and cloud formation
   A. Condensation
      1. Water vapor in air changes to liquid
         a. Dew
         b. Fog
         c. Clouds
      2. Water vapor requires a surface to condense on
         a. Possible condensation surfaces on the ground are grass, car window, etc.
         b. Possible condensation surfaces in the atmosphere
            1. Particulate matter
            2. Condensation nuclei
               a. Dust, smoke, etc.
               b. Ocean salt—absorbs water (hygroscopic nuclei)
   B. Clouds
      1. Made of millions and millions of
         a. Minute water droplets, or
         b. Tiny crystals of ice
      2. Classification based on
         a. Form (three basic forms)
            1. Cirrus—high, white, thin

2. Cumulus
    a. Globular cloud masses
    b. Often associated with fair weather
3. Stratus
    a. Sheets or layers
    b. Cover much or all of sky

b. Height
1. High clouds
    a. Above 6000 meters
    b. Types
        1. Cirrus
        2. Cirrostratus
        3. Cirrocumulus
2. Middle clouds
    a. 2000 to 6000 meters
    b. Types (alto as part of name)
        1. Altocumulus
        2. Altostratus
3. Low clouds
    a. Below 2000 meters
    b. Types
        1. Stratus
        2. Stratocumulus
        3. Nimbostratus (nimbus means "rainy")
4. Clouds of vertical development
    a. From low to high altitudes
    b. Cumulonimbus
        1. Rain showers
        2. Thunderstorms

VI. Fog
- A. Considered an atmospheric hazard
- B. Cloud with base at or near ground
- C. Formation
    1. Radiation cooling, or
    2. Movement of air over cold surface
- D. Types of fog
    1. Fogs caused by cooling
        a. Advection fog—warm, moist air over cool surface
        b. Radiation fog
            1. Earth's surface cools rapidly
            2. Cool, clear, calm nights
        c. Upslope fog
            1. Humid air moves up a slope
            2. Adiabatic cooling
    2. Fogs caused by evaporation
        a. Steam fog
            1. Cool air moves over warm water and moisture is added to the ai
            2. Water has steaming appearance
        b. Frontal fog, or precipitation fog
            1. Frontal wedging—warm air lifted over colder air
            2. Rain evaporates to form fog

VII. Precipitation
- A. Cloud droplets
    1. Less than 10 micrometers in diameter
    2. Fall incredibly slow
- B. Formation of precipitation
    1. Ice crystal process
        a. Temperature in cloud is below freezing
        b. Ice crystals collect water vapor
        c. Form large snowflakes
            1. Fall to ground as snow, or
            2. Melt on descent and form rain
    2. Collision-coalescence process
        a. Warm clouds
        b. Large hygroscopic condensation nuclei
        c. Large droplets form
        d. Collide with other droplets on descent
- C. Forms of precipitation
    1. Rain and drizzle
        a. Rain—droplets at least 0.5 mm diameter
        b. Drizzle—droplets less than 0.5 mm diameter
    2. Snow—ice crystals, or aggregates of crystals
    3. Sleet and glaze
        a. Sleet
            1. Wintertime phenomenon
            2. Small particles of ice
            3. Occurs when
                a. Warmer air overlies colder air
                b. Rain freezes as it falls
        b. Glaze, or freezing rain—impact with solid causes freezing
        c. Hail
            1. Hard rounded pellets
                a. Concentric shells
                b. Most diameters from 1 to 5 cm
            2. Formation
                a. In large cumulonimbus clouds with violent up- and downdrafts

b. Layers of freezing rain from up-and-down trips in cloud
c. Falls to ground when too heavy
d. Rime
1. Forms on cold surfaces
2. Freezing of
a. Supercooled fog, or
b. Cloud droplets

D. Measuring precipitation
1. Rain
1. Uses funnel to collect rain
a. Easiest form to measure
b. Measuring instrument standard rain gauge
2. Cylindrical measuring tube
a. Centimeters or inches
b. Recording gauge
2. Snow
a. Two measurements
1. Depth
2. Water equivalent
a. General ratio—10 snow units to 1 water unit
b. Varies widely

Vocabulary Review

Choosing from the list of key terms, furnish the most appropriate response for the following statements.

1. The general term for the amount of water vapor in air is ____________.

2. The process whereby water vapor changes to the liquid state is called ____________.

3. ____________ is the ratio of the air's actual water vapor content to its potential water vapor capacity at a given temperature.

4. ____________ results when elevated terrains, such as mountains, act as barriers to flowing air.

5. The part of total atmospheric pressure attributable to water vapor content is called the ________.

6. A(n) ____________ results when air is compressed or allowed to expand.

7. ____________ refers to the energy stored or released during a change of state.

8. A(n) ____________ is best described as visible aggregates of minute water droplets or tiny crystals of ice suspended in the air.

9. Frozen or semifrozen rain formed when raindrops freeze as they pass through a layer of cold air is called ____________.

10. The term ____________ refers to the maximum possible quantity of water vapor that the air can hold at any given temperature and pressure.

11. The rate of adiabatic temperature change in saturated air is the ____________.

12. ____________ is the conversion of a solid directly to a gas, without passing through the liquid state.

13. Sheets or layers of clouds that cover much or all of the sky are called ____________ clouds.

14. The weight of water vapor per weight of a chosen mass of air (e.g., grams per kilogram), including the water vapor, is called ____________.

Key Terms

Page numbers shown in ( ) refer to the textbook page where the term first appears.

adiabatic temperature change (p. 306)
advection fog (p. 314)
cirrus (p. 313)
cloud (p. 313)
condensation (p. 301)
condensation nuclei (p. 312)
convergence (p. 311)
cumulus (p. 313)
deposition (p. 301)
dew point (p. 303)
dry adiabatic rate (p. 307)
evaporation (p. 301)
fog (p. 314)
freezing (p. 301)
frontal fog (p. 315)
frontal wedging (p. 311)
glaze (p. 320)
hail (p. 320)
humidity (p. 301)
hygrometer (p. 303)
hygroscopic nuclei (p. 312)
latent heat (p. 300)
melting (p. 301)
orographic lifting (p. 310)
precipitation fog (p. 315)
radiation fog (p. 314)
rain (p. 319)
rainshadow desert (p. 311)
relative humidity (p. 302)
rime (p. 321)
saturation (p. 301)
sleet (p. 320)
snow (p. 320)
specific humidity (p. 302)
steam fog (p. 315)
stratus (p. 313)
sublimation (p. 301)
upslope fog (p. 314)
vapor pressure (p. 302)
wet adiabatic rate (p. 307)

15. ___________ occurs when cool air acts as a barrier over which warmer, less dense air rises.

16. In the atmosphere, tiny bits of particulate matter, known as ___________, serve as surfaces for water vapor condensation.

17. The process by which a solid is changed to a liquid is referred to as ___________.

18. In meteorology, the term ___________ is restricted to drops of water that fall from a cloud and have a diameter of at least 0.5 millimeter.

19. The temperature to which air would have to be cooled to reach saturation is the ___________.

20. ___________ is a deposit of ice crystals formed by the freezing of supercooled fog or cloud droplets on objects whose surface temperature is below freezing.

21. The conversion of a vapor directly to a solid is called ___________.

22. Nearly spherical ice pellets which have concentric layers and are formed by the successive freezing of layers of water are referred to as ___________.

23. The adiabatic rate of cooling or heating which applies only to unsaturated air is the __________.

24. Water absorbing particles, such as salt, are termed ___________.

25. The process of converting a liquid to a gas is called ___________.

26. ___________ is a type of fog that forms on cool, clear, calm nights, when Earth's surface cools rapidly by radiation.

27. A cloud with its base at or very near Earth's surface is referred to as ___________.

28. A dry area on the lee side of a mountain that forms as air descends and is warmed by compression, is called a(n) ___________.

29. The ___________, an instrument used to measure relative humidity, consists of two identical thermometers mounted side by side.

30. Whenever air masses flow together, ___________ is said to occur.

Comprehensive Review

1. Using Figure 12.1, select the letter that indicates each of the following processes.

   a) Freezing: ____

   b) Evaporation: ____

   c) Deposition: ____

   d) Melting: ____

   e) Sublimation: ___

   f) Condensation: ____

2. List two different ways that the relative humidity of a parcel of air can be increased.

   1)

   2)

3. Using Table 12.1 in the textbook, determine the relative humidity of a kilogram of air that has the following temperature and water vapor content.

   a) Temperature: 25°C; Water vapor content: 5 grams/kilogram

   b) Temperature: 41°F; Water vapor content: 3 grams/kilogram

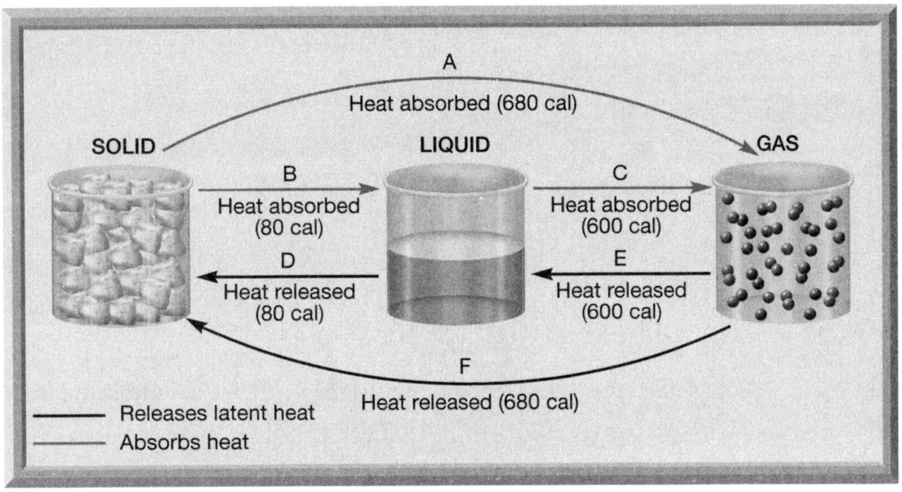

Figure 12.1

4. Using Table 12.1 in the textbook, determine the dew point of a kilogram of air that has the following temperature and relative humidity.

   a) Temperature: 25°C; Relative humidity: 50%

   b) Temperature: 59°F; Relative humidity: 20%

5. Explain the principle of the psychrometer for measuring relative humidity.

6. Using Table 12.2 in the textbook, determine the relative humidity of a parcel of air that yielded the following psychrometer readings.

   a) Dry-bulb temperature: 24°C; Wet-bulb temperature: 16°C

7. Why is the wet adiabatic rate of cooling of air less than the dry adiabatic rate?

8. What is the difference between stable and unstable air?

9. Briefly describe each of the following processes that lift air.

   a) Orographic lifting:

   b) Frontal wedging:

   c) Convergence:

10. What two criteria are used as the basis for cloud classification?

    1)

    2)

11. Using Figure 12.2, select the letter of the photograph that illustrates the following cloud types.

    a) Cumulus: ____ b) Cirrus: ____

12. Briefly describe the circumstances responsible for the formation of each of the following types of fog.

    a) Steam fog:

    b) Frontal fog:

13. List and briefly describe the two mechanisms that have been proposed to explain how precipitation forms.

    1)

    2)

14. Briefly describe the circumstances that result in the formation of each of the following forms of precipitation.

    a) Sleet:

    b) Hail:

A.

B.

Figure 12.2

Practice Test

Multiple choice. Choose the best answer for the following multiple choice questions.

1. The ratio of the air's water vapor content to its capacity at that same temperature is the ____________.
   a) vapor pressure c) wet adiabatic rate e) dry adiabatic rate
   b) specific humidity d) relative humidity

2. The term ____________ means "rainy cloud."
   a) ferrous c) stratus e) cumulus
   b) nimbus d) cirrus

3. Which one of the following changes from one state of matter to another at the temperatures and pressures experienced at Earth's surface?
   a) oxygen c) carbon dioxide e) methane
   b) water d) nitrogen

4. Which of the following is NOT a necessary condition for condensation?
   a) high altitude
   b) dew-point temperature reached
   c) surfaces
   d) saturation
   e) water vapor

5. Which one of the following processes involves the greatest quantity of heat energy?
   a) melting c) freezing e) condensation
   b) sublimation d) evaporation

6. Which one of the following is NOT a process that lifts air?
   a) convergence b) orographic lifting c) divergence d) frontal wedging

7. Which form of precipitation is likely to occur when a layer of warm air with temperatures above freezing overlies a subfreezing layer near the ground?
   a) rain c) snow e) drizzle
   b) sleet d) hail

8. Air that has reached its water vapor capacity is said to be ____________.
   a) dry c) soaked e) saturated
   b) unstable d) stable

9. Which clouds are best described as sheets or layers that cover much or all of the sky?
   a) cirrus c) stratus e) cumulonimbus
   b) cumulus d) altocumulus

10. The temperature to which air would have to be cooled to reach saturation is called the ____________ point.
    a) vapor c) adiabatic e) critical
    b) dew d) sublimation

11. When the environmental lapse rate is less than the dry adiabatic rate, a parcel of air will be ____________.
    a) stable
    b) unstable

12. Which one of the following refers to the energy that is stored or released during a change of state of water?
    a) caloric heat
    b) ultraviolet heat
    c) latent heat
    d) geothermal heat
    e) evaporation heat

13. The process where cool air acts as a barrier over which warmer, less dense air rises is called ____________.
    a) divergence
    b) frontal wedging
    c) orographic lifting
    d) collision
    e) subduction

14. Relative humidity is typically highest at ____________.
    a) sunrise
    b) noon
    c) late afternoon
    d) early evening
    e) midnight

15. With which cloud type is hail most associated?
    a) cirrus
    b) stratus
    c) cumulonimbus
    d) nimbostratus
    e) cumulus

16. The wet adiabatic rate of cooling is less than the dry rate because ____________.
    a) wet air is unsaturated
    b) of the release of latent heat
    c) dry air is less dense
    d) wet air is more common
    e) of the dew point

17. The conditions that favor the formation of radiation fog are ____________.
    a) warm, cloudy, calm nights
    b) cool, cloudy, windy nights
    c) warm, clear, windy nights
    d) cool, clear, calm nights
    e) warm, cloudy, windy nights

18. The process responsible for deposits of white frost or hoar frost is ____________.
    a) sublimation
    b) melting
    c) freezing
    d) evaporation
    e) deposition

19. When the environmental lapse rate is greater than the dry adiabatic rate, a parcel of air will be ____________.
    a) stable
    b) unstable

20. Compared to clouds, fogs are ____________.
    a) colder
    b) thicker
    c) of a different composition
    d) at lower altitudes
    e) drier

21. Which clouds are high, white, and thin?
    a) cirrus
    b) cumulus
    c) stratus
    d) nimbostratus
    e) cumulonimbus

22. When air expands or contracts it will experience ____________ temperature changes.
    a) volume
    b) radiation
    c) external
    d) adiabatic
    e) unnecessary

23. Which one of the following is NOT a process that has been proposed to explain the formation of precipitation?
    a) atmospheric subduction process
    b) collision-coalescence process
    c) ice crystal process

24. The conversion of a solid directly to a gas, without passing through the liquid state, is called ___________.
    a) sublimation c) freezing e) deposition
    b) melting d) evaporation

25. Weather-producing fronts are parts of the storm systems called ___________.
    a) mid-latitude cyclones c) tornadoes e) typhoons
    b) hurricanes d) tropical storms

26. Which one of the following is NOT produced by condensation?
    a) dew b) smog c) fog d) clouds

27. Which one of the following is a deposit of ice crystals formed by the freezing of supercooled fog or cloud droplets on objects whose surface temperature is below freezing?
    a) hail c) snow e) glaze
    b) sleet d) rime

28. The dry adiabatic rate is ___________.
    a) 3.0°C/100 meters d) 1°C/1000 meters
    b) 0.5°C/meter e) 0.5°C/1000 meters
    c) 1°C/100 meters

29. That part of total atmospheric pressure that can be attributed to the water vapor content is called ___________ pressure.
    a) moisture c) vapor e) water
    b) isostatic d) gas

30. Which one of the following is NOT a type of fog?
    a) white fog c) upslope fog e) frontal fog
    b) radiation fog d) steam fog

True/false. For the following true/false questions, if a statement is not completely true, mark it false. For each false statement, change the ***italicized*** *word to correct the statement.*

1. ___ At the *dew-point* temperature, the air is both saturated and has a 100% relative humidity.
2. ___ A *dyne* is the amount of heat required to raise the temperature of 1 gram of water 1°C.
3. ___ Clouds are classified on the basis of their form and *color.*
4. ___ When water vapor condenses, it *releases* heat energy.
5. ___ When air expands, it will *warm.*
6. ___ When air masses flow together, *convergence* is said to occur.

7. ___ Assuming equal volumes, a *greater* quantity of water vapor is required to saturate warm air than cold air.

8. ___ Clouds associated with *unstable* air are towering and usually accompanied by heavy precipitation.

9. ___ The principle of the psychrometer relies on the fact that the amount of cooling of the *dry-bulb* thermometer is directly related to the dryness of the air.

10. ___ Above the ground, tiny bits of particulate matter serve as *condensation* nuclei.

11. ___ The water vapor capacity of air is *temperature*-dependent.

12. ___ The *dry* adiabatic rate varies from about 5°C/1000 meters in moist air to 9°C/1000 meters in dry air.

13. ___ When the water vapor content remains constant, a(n) *increase* in temperature results in an increase in relative humidity.

14. ___ A raindrop large enough to fall to the ground contains roughly one *million* times more water than a single cloud droplet.

15. ___ The energy absorbed by water molecules during evaporation is referred to as latent heat of *fusion*.

16. ___ When the surface temperature is about *four* degrees Celsius or higher, snowflakes usually melt before they reach the ground and continue their descent as rain.

17. ___ *Sleet* is defined as a cloud with its base at or very near the ground.

18. ___ Although highly variable, a general snow/water ratio of *ten* units of snow to one unit of water is often used when exact information is not available.

Written questions

1. Describe how clouds are classified.

2. Explain why air cools as it rises through the atmosphere.

3. By comparing the environmental lapse rate to the adiabatic rate, describe when air will be stable and when it will be unstable.

4. Of the three mechanisms that lift air, which is most prevalent in your location? Explain how this mechanism is responsible for the precipitation of your area.

13

The Atmosphere in Motion

The Atmosphere in Motion opens with a description of the units and instruments used for measuring atmospheric pressure. A definition of wind is followed by an analysis of the factors that affect it—pressure gradient force, Coriolis effect, and friction. A discussion of cyclones and anticyclones includes associated movements of air and weather patterns. Also presented are the general, global patterns of pressure and wind. A more detailed discussion of atmospheric circulation in the mid-latitudes is followed by descriptions of several local winds. Wind measurement and instruments are briefly mentioned at the end of the chapter.

Learning Objectives

After reading, studying, and discussing this chapter, you should be able to:

- Describe air pressure, how it is measured, and how it changes with altitude.
- Explain how the pressure gradient force, Coriolis effect, and friction influence wind.
- Describe the movements of air associated with the two types of pressure centers.
- Describe the idealized global patterns of pressure and wind.
- Discuss the general atmospheric circulation in the mid-latitudes.
- List the names and causes of the major local winds.

Chapter Review

- *Air has weight:* at sea level it exerts a pressure of 1 kilogram per square centimeter (14.7 pounds per square inch). *Air pressure* is the force exerted by the weight of air above. With increasing altitude there is less air above to exert a force, and thus air pressure decreases with altitude, rapidly at first, then much more slowly. The unit used by meteorologists to measure atmospheric pressure is the *millibar*. *Standard sea level pressure* is expressed as 1013.2 millibars. *Isobars* are lines on a weather map that connect places of equal air pressure.

- A *mercury barometer* measures air pressure using a column of mercury in a glass tube that is sealed at one end and inverted in a dish of mercury. As air pressure increases, the mercury in the tube rises; conversely, when air pressure decreases, so does the height of the column of mercury. A mercury barometer measures atmospheric pressure in "*inches of mercury*," the height of the column of mercury in the barometer. Standard atmospheric pressure at sea level equals 29.92 inches of mercury. *Aneroid* ("without liquid") *barometers* consist of partially-evacuated metal chambers that compress as air pressure increases and expand as pressure decreases.

- *Wind* is the horizontal flow of air from areas of higher pressure toward areas of lower pressure. Winds are controlled by the following combination of forces: 1) the *pressure gradient force* (amount of pressure change over a given distance), 2) *Coriolis effect* (deflective effect of Earth's rotation—to the right in the Northern Hemisphere and to the left in the Southern Hemisphere), 3) *friction* with Earth's surface (slows the movement of air and alters wind direction), and 4) the *tendency* of a moving object *to continue moving in a straight line.*

• The two types of pressure centers are 1) *cyclones*, or *lows* (centers of low pressure), and 2) *anticyclones*, or *highs* (high-pressure centers). In the Northern Hemisphere, winds around a low (cyclone) are counterclockwise and inward. Around a high (anticyclone) they are clockwise and outward. In the Southern Hemisphere, the Coriolis effect causes winds to be clockwise around a low and counterclockwise around a high. Since air rises and cools adiabatically in a low pressure center, cloudy conditions and precipitation are often associated with their passage. In a high pressure center, descending air is compressed and warmed; therefore, cloud formation and precipitation are unlikely in an anticyclone, and "fair" weather is usually expected.

• Earth's *global pressure zones* include the *equatorial low*, *subtropical high*, *subpolar low*, and *polar high*. The *global surface winds* associated with these pressure zones are the *trade winds*, *westerlies*, and *polar easterlies*.

• Particularly in the Northern Hemisphere, large seasonal temperature differences over continents disrupt the idealized, or zonal, global patterns of pressure and wind. In winter, large, cold landmasses develop a seasonal high-pressure system from which surface air flow is directed off the land. In summer, landmasses are heated and a low-pressure system develops over them, which permits air to flow onto the land. These seasonal changes in wind direction are known as *monsoons*.

• In the middle latitudes, between 30 and 60 degrees latitude, the general west-to-east flow of the westerlies is interrupted by the migration of cyclones and anticyclones. The paths taken by these cyclonic and anticyclonic systems is closely correlated to upper-level air flow and the polar *jet stream*. The average position of the polar jet stream, and hence the paths followed by cyclones, migrates southward with the approach of winter and northward as summer nears.

• *Local winds* are small-scale winds produced by a locally generated pressure gradient. Local winds include *sea* and *land breezes* (formed along a coast because of daily pressure differences over land and water), *valley* and *mountain breezes* (daily wind similar to sea and land breezes except in a mountainous area where the air along slopes heats differently than the air at the same elevation over the valley floor), *chinook* and *Santa Ana winds* (warm, dry winds created when air descends the leeward side of a mountain and warms by compression).

• The two basic wind measurements are *direction* and *speed*. Winds are always labeled by the direction *from* which they blow. Wind direction is measured with a *wind vane* and wind speed is measured using a *cup anemometer*.

Chapter Outline

I. Atmospheric pressure
    A. Force exerted by weight of air above
    B. Weight of the air at sea level
        1. 14.7 pounds per square inch
        2. 1 kilogram per square centimeter
    C. Decreases with increasing altitude
    D. Units of measurement
        1. Millibar (mb)—standard sea level pressure is 1013.2 mb
        2. Inches of mercury standard sea level pressure is 29.92 inches of mercury
    E. Instruments for measuring
        1. Barometer
            a. Mercury barometer
            b. Aneroid barometer
        2. Barograph (continuously records)

II. Wind
  A. Horizontal movement of air
    1. Out of areas of high pressure
    2. Into areas of low pressure
  B. Controls of wind
    1. Pressure gradient force
      a. Isobars—lines of equal air pressure
      b. Pressure gradient—pressure change over distance
    2. Coriolis effect
      a. Apparent deflection in wind direction due to Earth's rotation
      b. Deflection
        1. To right in Northern Hemisphere
        2. To left in Southern Hemisphere
    3. Friction with Earth's surface
      a. Only important near the surface
      b. Acts to slow movement
  C. Upper air winds
    1. Generally blow parallel to isobars (geostrophic winds)
    2. Jet stream
      a. "River" of air
      b. High altitude
      c. High velocity (120-240 km/hour)
III. Cyclones and anticyclones
  A. Cyclone
    1. Center of low pressure
    2. Pressure decreases toward center
    3. Winds associated with
      a. In Northern Hemisphere
        1. Inward (convergence)
        2. Counterclockwise
      b. In Southern Hemisphere
        1. Inward (convergence)
        2. Clockwise
    4. Associated with rising air
    5. Often bring clouds and precipitation
  B. Anticyclone
    1. Center of high pressure
    2. Pressure increases toward center
    3. Winds associated with
      a. In Northern Hemisphere
        1. Outward (divergence)
        2. Clockwise
      b. In Southern Hemisphere
        1. Outward (divergence)
        2. Counterclockwise
    4. Associated with subsiding air
    5. Usually bring "fair" weather
IV. General atmospheric circulation
  A. Underlying cause is unequal surface heating—on rotating Earth three pairs of cells redistribute heat
  B. Idealized global circulation
    1. Equatorial low pressure zone
      a. Rising air
      b. Abundant precipitation
    2. Subtropical high pressure zone
      a. Subsiding, stable, dry air
      b.Near 30 degrees latitude
      c. Location of great deserts
      d. Air traveling equatorward from subtropical high produces the trade winds
      e. Air traveling poleward from subtropical high produces the westerly winds
    3. Subpolar low pressure zone
      a. Warm and cool winds interact
      b. Polar front area of storms
    4. Polar high pressure zone
      a. Cold, subsiding air
      b. Spreads equatorward and produces polar easterly winds
      c. Polar easterlies collide with westerlies along polar the front
  C. Influence of continents
    1. Seasonal temperature differences disrupt
      a. Global pressure patterns
      b. Global wind patterns
    2. Most obvious in Northern Hemisphere
    3. Monsoon
      a. Seasonal change in wind direction
      b. Occur over continents
        1. During warm months
          a. Air flows onto land
          b. Warm, moist air from ocean
        2. Winter months
          a. Air flows off the land
          b. Dry, continental air
V. Circulation in the mid-latitudes
  A. Complex
  B. In zone of westerlies
  C. Air flow interrupted by cyclones
    1. Cells move west to east in N. Hemisphere

2. Create anticyclonic and cyclonic flow
3. Paths associated with upper-level airflow

VI. Local winds
   A. Produced from temperature differences
   B. Small scale winds
   C. Types
      1. Sea and land breezes
      2. Valley and mountain breezes
      3. Chinook and Santa Ana winds

VII. Wind measurement
   A. Two basic measurements
      1. Direction
      2. Speed
   B. Direction
      1. Winds labeled from where they originate (e.g., North wind—blows from north toward south)
      2. Instrument for measuring wind direction is the wind vane
      3. Direction indicated by either
         a. Compass points (N, NE, E, etc.)
         b. Scale of 0° to 360°
      4. Prevailing wind—wind more often from one direction
   C. Speed—often measured with cup anemometer
   D. Changes in wind direction
      1. Associated with locations of
         a. Cyclones
         b. Anticyclones
      2. Often bring changes in
         a. Temperature
         b. Moisture conditions

Vocabulary Review

Choosing from the list of key terms, furnish the most appropriate response for the following statements.

Key Terms

Page numbers shown in ( ) refer to the textbook page where the term first appears.

air pressure (p. 328)
aneroid barometer (p. 328)
anticyclone (high) (p. 333)
barograph (p. 328)
barometric tendency (p. 335)
chinook (p. 340)
convergence (p. 334)
Coriolis effect (p. 330)
cup anemometer (p. 341)
cyclone (low) (p. 333)
divergence (p. 334)
equatorial low (p. 336)
geostrophic wind (p. 331)
isobar (p. 329)
jet stream (p. 331)
land breeze (p. 339)
mercury barometer (p. 328)
millibar (p. 328)
monsoon (p. 337)
mountain breeze (p. 340)
polar easterlies (p. 336)
polar front (p. 336
polar high (p. 336)
pressure gradient (p. 329)
pressure tendency (p. 335)
prevailing wind (p. 341)
Santa Ana (p. 340)
sea breeze (p. 339)
subpolar low (p. 336)
subtropical high (p. 336)
trade winds (p. 336)
valley breeze (p. 340)
westerlies (p. 336)
wind (p. 329)
wind vane (p. 340)

1. The stormy belt separating the westerlies from the polar easterlies is known as the ____________.

2. A(n) ____________ is a line on a weather map that connects places of equal air pressure.

3. A(n) ____________ is a seasonal reversal of wind direction associated with large continents, especially Asia.

4. The deflective force of Earth's rotation on all free-moving objects is called the ____________.

5. Air that flows horizontally with respect to Earth's surface is referred to as ___________.

6. ___________, a useful aid in short-range weather prediction, refers to the nature of the change in atmospheric pressure over the past several hours.

7. The ___________ is an instrument used for measuring air pressure that consists of evacuated metal chambers that change shape as pressure changes.

8. When the wind consistently blows more often from one direction than from any other, it is termed a(n) ___________.

9. The instrument most commonly used to determine wind direction is the ___________.

10. A center of low atmospheric pressure is called a(n) ___________.

11. ___________ is the condition that exists when the distribution of winds within a given area results in a net horizontal inflow of air into the area.

12. A(n) ___________ is a local wind blowing from land toward the water during the night in coastal areas.

13. The ___________ is a belt of low pressure lying near the equator and between the subtropical highs.

14. A wind blowing down the leeward side of a mountain and warming by compression is called a(n) ___________.

15. A(n) ___________ is a local wind blowing from the sea during the afternoon in coastal areas.

16. The ___________ is an instrument used to measure wind speed.

17. The amount of pressure change over a given distance is referred to as the ___________.

18. A(n) ___________ is a swift (120 to 240 km/hour), high-altitude wind.

19. A center of high atmospheric pressure is called a(n) ___________.

20. The ___________ is the pressure zone located at about the latitude of the Arctic and Antarctic circles.

21. The ___________ are global winds that blow from the polar high toward the subpolar low.

22. A(n) ___________ is an instrument that continuously records air pressure changes.

23. ___________ is the condition that exists when the distribution of winds within a given area results in a net horizontal outflow of air from the region.

24. The ___________ is a region of several semipermanent anticyclonic centers characterized by subsidence and divergence located roughly between latitudes 25 and 35 degrees.

25. The ___________ are the dominant west-to-east winds that characterize the regions on the poleward sides of the subtropical highs.

26. The ___________ are wind belts located on the equatorward sides of the subtropical highs.

Comprehensive Review

1. Which element of weather is measured by each of the following instruments? Briefly describe the principle of each instrument.
   a) Mercury barometer:

   b) Wind vane:

   c) Aneroid barometer:

   d) Cup anemometer:

   e) Barograph:

2. List and describe the three forces that control the wind.

   1)

   2)

   3)

3. On Figure 13.1, complete each of the map-views of the four pressure center diagrams by labeling the isobars with appropriate pressures and drawing several wind arrows to indicate surface air movement.

4. Referring to Figure 13.1, describe the general surface air flow associated with each of the following pressure systems.
   a) Northern Hemisphere low (cyclone):

   b) Southern Hemisphere high (anticyclone):

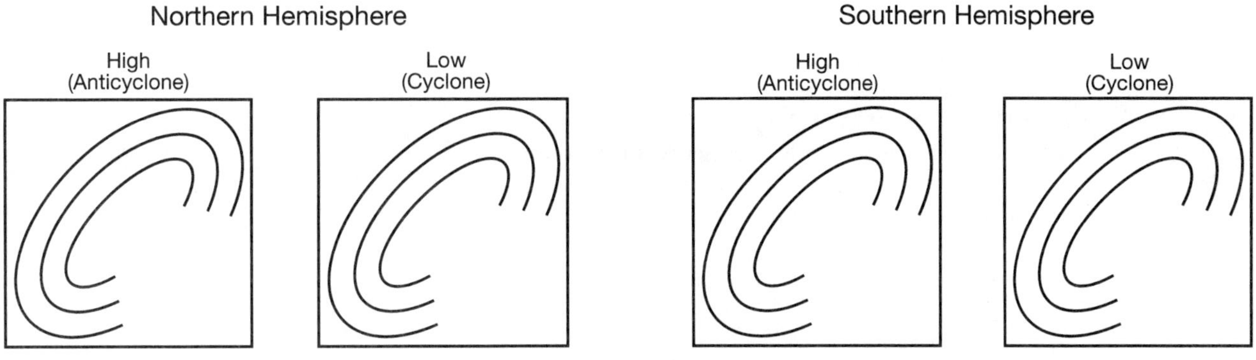

Figure 13.1

5. Figure 13.2 illustrates side-views of the air movements in a cyclone and an anticyclone. Using Figure 13.2, select the letter of the diagram that illustrates each of the following:
   a) Cyclone: ____
   b) Convergence aloft: ____
   c) Surface convergence: ____
   d) Anticyclone: ____
   e) Divergence aloft: ____

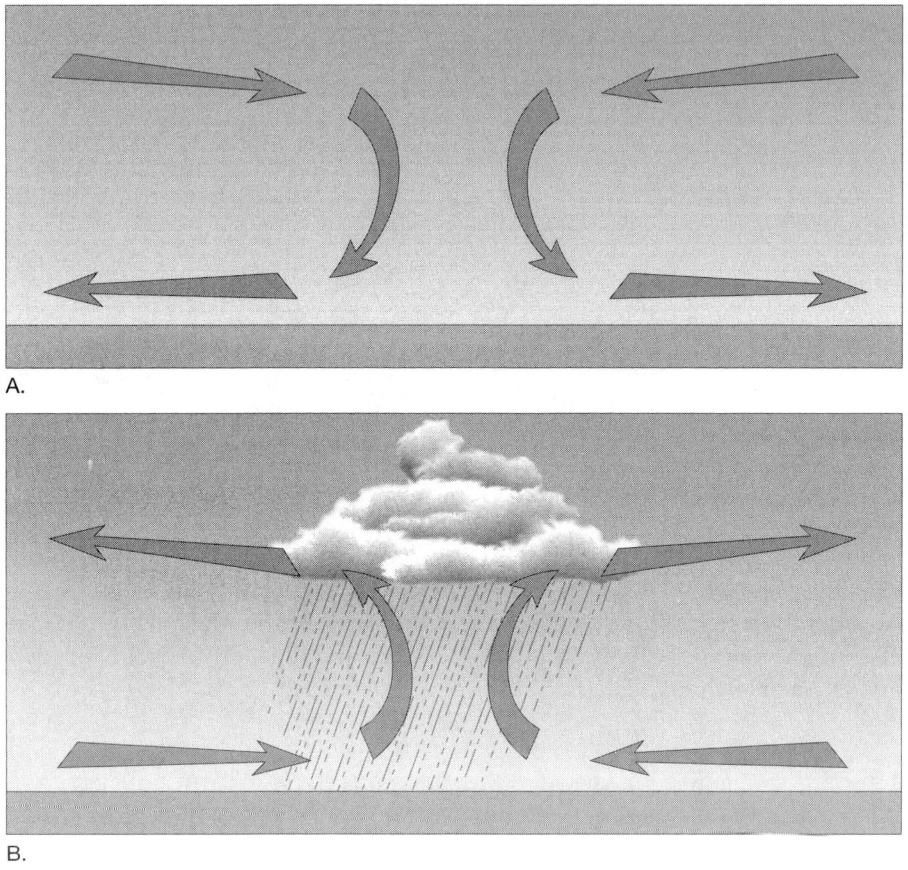

Figure 13.2

6. Why is "fair" weather usually associated with the approach of a high pressure system (anticyclone)?

7. On Figure 13.3, write the names of the global pressure zones indicated by letters A through E; and the wind belts indicated by letters F through K.

8. Write the name of the global pressure zone best described by each of the following statements.
   a) Warm and cold winds interact to produce stormy weather:

   b) Subsiding, dry air with extensive arid and semiarid regions:

   c) Warm, rising air marked by abundant precipitation:

   d) Cold, subsiding, polar air:

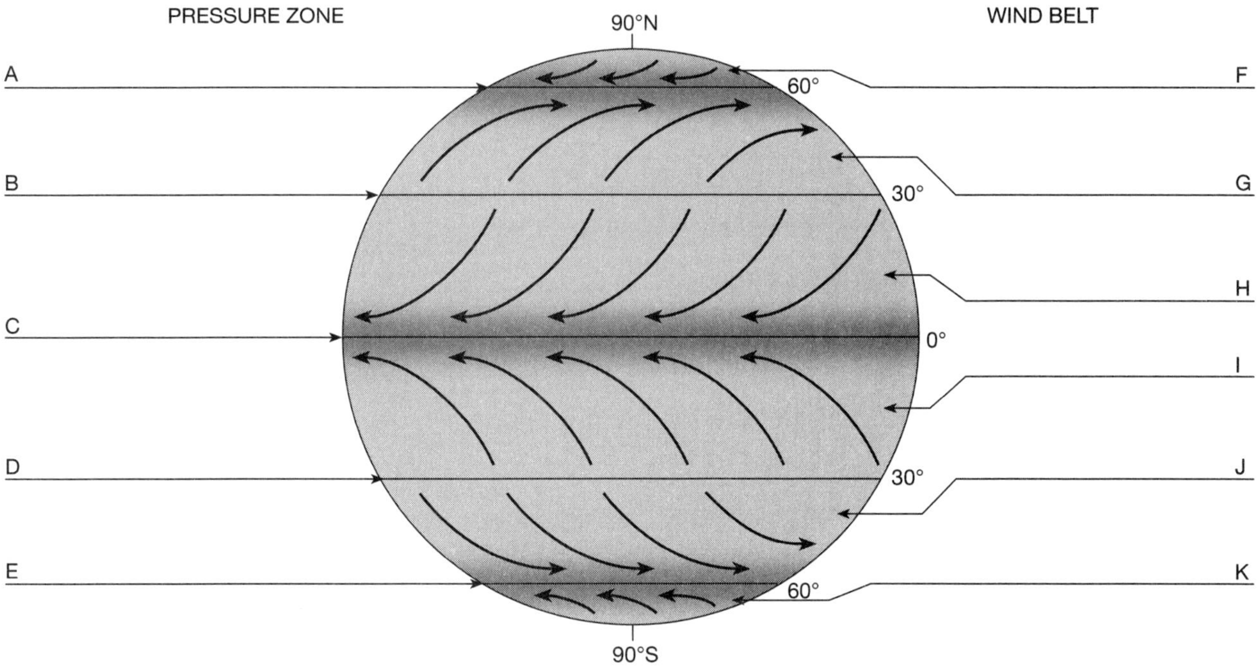

Figure 13.3

9. Briefly describe each of the following local winds.
   a) Chinook:

   b) Sea breeze:

10. If a wind is indicated with the following degree, from what compass direction is it blowing?
    a) 45 degrees:

    b) 270 degrees:

    c) 180 degrees:

Practice Test

Multiple choice. Choose the best answer for the following multiple choice questions.

1. The only truly continuous pressure belt on Earth is the ___________.
   a) Northern Hemisphere subtropical high
   b) equatorial low
   c) Southern Hemisphere subtropical high
   d) Southern Hemisphere subpolar low
   e) Northern Hemisphere subpolar low

2. The force exerted by the weight of the air above is called ____________.
   a) air pressure c) the Coriolis effect e) divergence
   b) convergence d) the pressure gradient

3. Which one of the following is NOT a force that controls wind?
   a) magnetic force c) pressure gradient force
   b) Coriolis effect d) friction

4. Variations in air pressure from place to place are the principle cause of ____________.
   a) snow c) wind e) hail
   b) rain d) clouds

5. Fair weather can usually be expected with the approach of a(n) ____________.
   a) cyclone b) anticyclone

6. Centers of low pressure are called ____________.
   a) anticyclones c) cyclones e) domes
   b) air masses d) jet streams

7. In the winter, large landmasses, particularly Asia, develop a seasonal ____________.
   a) high-pressure system c) low-pressure system e) cyclonic circulation
   b) trade wind system d) chinook

8. Wind with a direction of 225° would be a ____________ wind.
   a) west c) south e) northwest
   b) northeast d) southwest

9. The general movement of low pressure centers across the United States is from ____________.
   a) north to south c) east to west e) northeast to southwest
   b) south to north d) west to east

10. Near the equator, rising air is associated with a pressure zone known as the ____________.
    a) equatorial high c) equatorial low e) subtropical low
    b) tropical high d) tropical low

11. Standard sea level pressure is ____________.
    a) 1020.6 millibars
    b) 15.8 pounds per square inch
    c) 29.92 inches of mercury
    d) 3.7 kilograms per square centimeter
    e) 100 kilopascals

12. A sea breeze is most intense ____________.
    a) during mid- to late afternoon
    b) in the late morning
    c) late in the evening
    d) at midnight
    e) at sunrise

13. High-altitude, high-velocity "rivers" of air are referred to as ____________.
    a) cyclones c) anticyclones e) jet streams
    b) tornadoes d) air currents

14. The deflection of wind due to the Coriolis effect is strongest at ___________.
    a) the equator c) midnight e) sunrise
    b) the mid-latitudes d) the poles

15. Seasonal changes in wind direction associated with large landmasses and adjacent water bodies are called ___________.
    a) chinooks c) jet streams e) monsoons
    b) geostrophic winds d) trade winds

16. Pressure data on a map are shown using lines that connect places of equal air pressure called ___________.
    a) isotherms c) aneroids e) isobars
    b) millibars d) kilograms

17. Air is subsiding in the center of a ___________.
    a) low pressure system c) surface convergence area e) chinook
    b) jet stream d) high pressure system

18. A ___________ is an instrument that continuously records pressure changes.
    a) mercury barometer c) aneroid barometer e) cup anemometer
    b) barograph d) sling psychrometer

19. The surface winds between the subtropical high and equatorial low pressure zones are the ___________.
    a) polar easterlies c) trade winds e) westerlies
    b) sea breezes d) monsoon winds

20. Which one of the following does NOT describe the surface air movement in a Northern Hemisphere low?
    a) inward b) counterclockwise c) net upward movement d) divergent

21. Winds that blow parallel to isobars are called ___________.
    a) monsoon winds c) sea breezes e) trade winds
    b) geostrophic winds d) chinooks

22. Chinooks are warm, dry winds that commonly occur on the ___________ slopes of the Rockies.
    a) northern b) eastern c) southern d) western

23. The westerlies and polar easterlies converge in a stormy region known as the ___________.
    a) polar front c) subtropical front e) polar low
    b) equatorial low d) subpolar high

24. In the mid-latitudes, the movement of cyclonic and anticyclonic systems is steered by the ___________.
    a) polar easterlies c) trade winds e) ocean circulation
    b) upper-level air flow d) pressure gradient

25. The surface winds between the subtropical high and subpolar low pressure zones are the ___________.
    a) polar easterlies c) trade winds e) westerlies
    b) sea breezes d) monsoon winds

True/false. For the following true/false questions, if a statement is not completely true, mark it false. For each false statement, change the ***italicized*** *word to correct the statement.*

1. ___ The ultimate driving force of wind is *solar* energy.
2. ___ The greater the pressure differences between two places, the *slower* the wind speed.
3. ___ Centers of high pressure are called *anticyclones.*
4. ___ In the Northern Hemisphere, the Coriolis effect causes the path of wind to be curved to the *left.*
5. ___ The *aneroid* barometer consists of metal chambers that change shape as air pressure changes.
6. ___ A *rising* barometric tendency often means that fair weather can be expected.
7. ___ Upper air flow is often nearly *parallel* to the isobars.
8. ___ Pressure decreases from the outer isobars toward the center in an *anticyclone.*
9. ___ In India, the winter monsoon is dominated by *wet* continental air.
10. ___ A steep pressure gradient indicates *strong* winds.
11. ___ A *low* pressure system often brings cloudiness and precipitation.
12. ___ As wind speed increases, deflection by the Coriolis effect becomes *less.*
13. ___ The pressure zone of subsiding, dry air which encircles the globe near 30 degrees latitude, north and south, is the *mid-latitude* high.
14. ___ Wind direction is modified by both the Coriolis effect and *friction.*
15. ___ In both hemispheres, an anticyclone is associated with *divergence* of the air aloft.
16. ___ In the Northern Hemisphere, more cyclones are generated during the *warmer* months when the temperature gradient is greatest.

Written questions

1.How does the Coriolis effect influence the movement of air?

2.Briefly describe the weather that is usually associated with 1) a cyclone and 2) an anticyclone.

3.Describe the motions of air at the surface and aloft in a Northern Hemisphere cyclone.

Weather Patterns and Severe Weather

14

Weather Patterns and Severe Weather begins with a discussion of air masses, their source regions, and a description of the weather associated with each air mass type. Following a detailed examination of warm fronts and cold fronts is an in-depth discussion of the evolution of the middle-latitude cyclone and its idealized weather patterns. The chapter concludes with investigations of thunderstorms, tornadoes, and hurricanes.

Learning Objectives

After reading, studying, and discussing this chapter, you should be able to:

- Explain what an air mass is.
- Describe how air masses are classified.
- Describe the general weather associated with each air mass type.
- Discuss the differences between warm fronts and cold fronts.
- Describe the primary mid-latitude weather producing systems.
- List the atmospheric conditions that produce thunderstorms, tornadoes, and hurricanes.

Chapter Review

- An *air mass* is a large body of air, usually 1600 kilometers (1000 miles) or more across, which is characterized by a *homogeneity of temperature and moisture* at any given altitude. When this air moves out of its region of origin, called the *source region,* it will carry these temperatures and moisture conditions elsewhere, perhaps eventually affecting a large portion of a continent.

- Air masses are classified according to 1) the nature of the surface in the source region and 2) the latitude of the source region. *Continental (c)* designates an air mass of land origin, with the air likely to be dry; whereas a *maritime (m)* air mass originates over water, and therefore will be humid. *Polar (P)* air masses originate in high latitudes and are cold. *Tropical (T)* air masses form in low latitudes and are warm. According to this classification scheme, the *four basic types of air masses* are *continental polar (cP), continental tropical (cT), maritime polar (mP),* and *maritime tropical (mT).* Continental polar (cP) and maritime tropical (mT) air masses influence the weather of North America most, especially east of the Rocky Mountains. Maritime tropical air is the source of much, if not most, of the precipitation received in the eastern two-thirds of the United States.

- *Fronts* are boundaries that separate air masses of different densities, one warmer and often higher in moisture content than the other. A *warm front* occurs when the surface position of the front moves so that warm air occupies territory formerly covered by cooler air. Along a warm front, a warm air mass overrides a retreating mass of cooler air. As the warm air ascends, it cools adiabatically to produce clouds and frequently, light-to-moderate precipitation over a large area. A *cold front* forms where cold air is actively advancing into a region occupied by warmer air. Cold fronts are about twice as steep and move more rapidly than warm fronts. Because of these two differences, precipitation along a cold front is more intense and of shorter duration than precipitation associated with a warm front.

• The primary weather producers in the middle latitudes are *large centers of low pressure* that generally travel from *west to east*, called *middle-latitude cyclones.* These *bearers of stormy weather*, which last from a few days to a week, have a *counterclockwise circulation* pattern in the Northern Hemisphere, with an *inward flow of air* toward their centers. Most middle-latitude cyclones have a *cold front and frequently a warm front* extending from the central areas of low pressure. *Convergence and forceful lifting along the fronts* initiate cloud development and frequently cause precipitation. As a middle-latitude cyclone with its associated fronts passes over a region, it often brings with it abrupt changes in the weather. The particular weather experienced by an area depends on the path of the cyclone.

• *Thunderstorms* are caused by the upward movement of warm, moist, unstable air, triggered by a number of different processes. They are associated with cumulonimbus clouds that generate heavy rainfall, thunder, lightning, and occasionally hail and tornadoes. *Tornadoes*, destructive, local storms of short duration, are violent windstorms associated with severe thunderstorms that take the form of a rotating column of air that extends downward from a cumulonimbus cloud. Tornadoes are most often spawned along the cold front of a middle-latitude cyclone, most frequently during the spring months. *Hurricanes*, the greatest storms on Earth, are tropical cyclones with wind speeds in excess of 119 kilometers (74 miles) per hour. These complex tropical disturbances develop over tropical ocean waters and are fueled by the latent heat liberated when huge quantities of water vapor condense. Hurricanes form most often in late summer when ocean-surface temperatures reach 27°C (80°F) or higher and thus are able to provide the necessary heat and moisture to the air. Hurricanes diminish in intensity whenever they 1) move over cool ocean water that cannot supply adequate heat and moisture, 2) move onto land, or 3) reach a location where large-scale flow aloft is unfavorable.

Chapter Outline

I. Air masses
  A. Characteristics
    1. Large body of air
      a. 1600 km (1000 mi.) or more across
      b. Several kilometers thick
    2. Similar temperature at any given altitude
    3. Similar moisture at any given altitude
    4. Move and affect large part of a continent
  B. Source regions—where air mass acquires its properties
  C. Classification of an air mass
    1. Two criteria used to classify
      a. By nature of surface in source region
        1. Continental (c)
          a. Form over land
          b. Likely to be dry
        2. Maritime (m)
          a. Originate over water
          b. Humid
      b. By source region
        1. Polar (P)
          a. High latitudes
          b. Cold
        2. Tropical (T)
          a. Low latitudes
          b. Warm
    2. Four basic types of air masses
      a. Continental polar (cP)
      b. Continental tropical (cT)
      c. Maritime polar (mP)
      d. Maritime tropical (mT)
  D. Air masses and weather
    1. cP and mT most important in North America
    2. North America (east of Rocky Mtns.)

a. Continental polar (cP)
1. From northern Canada, interior Alaska
a. Winter—brings cold, dry air
b. Summer—brings cool relief
2. Responsible for lake-effect snows
a. cP air mass crosses Great Lakes
b. Air picks up moisture from lakes
c. Snow on leeward shores of lakes
b. Maritime tropical (mT)
1. From Gulf of Mexico, Atlantic Ocean
2. Warm, moist, unstable air
3. Brings precipitation to eastern U.S.
3. Continental tropical (cT)
a. Southwest and Mexico
b. Hot, dry
c. Seldom important outside source region
4. Maritime polar (mP)
a. Precipitation to western mountains
b. Occasional influence in northeastern U.S.
1. "Northeaster" in New England
2. Cold temperatures and snow

II. Fronts
A. Boundary that separates air masses
1. Air masses retain their identities
2. Warmer, less dense air forced aloft
3. Cooler, denser air acts as wedge
B. Types of fronts
1. Warm front
a. Warm air replaces cooler air
b. Shown on map with semicircles
c. Small slope (1:200)
d. Clouds become lower as front nears
e. Slow rate of advance
f. Light-to-moderate precipitation
g. Gradual temperature increase with passage
2. Cold front
a. Cold air replaces warm air
b. Shown on map with triangles
c. Twice as steep as warm fronts (1:100)
d. Advances faster than warm front
e. Weather more violent than warm front
1. Intensity of precipitation is greater
2. Duration of precipitation is shorter
f. Weather behind front is dominated by
1. Cold air mass
2. Subsiding air
3. Clearing conditions

III. Middle-latitude cyclones (N. Hemisphere)
A. Primary weather producer in middle-latitudes
B. General characteristics
1. Large center of low pressure
a. Counterclockwise air circulation
b. Air flows inward toward center
2. Travel west to east, guided by westerlies
3. Last a few days to more than a week
4. Extending from center of the low are
a. Cold front
b. Frequently a warm front
5. Convergence and forceful lifting cause
a. Cloud development
b. Abundant precipitation
C. Idealized weather
1. Move eastward across United States
a. First signs of approach are in the west
b. Require two to four days to pass
2. Largest weather contrasts occur in spring
3. Changes in weather associated with passage
a. Changes depend on path of storm
b. Weather associated with fronts
1. Warm front
a. Clouds become lower, thicker
b. Light precipitation
1. Perhaps over large area
2. Perhaps prolonged duration
c. After passage
1. Winds more southerly
2. Warmer temperature (mT air mass)
2. Cold front
a. Wall of dark clouds
b. Heavy precipitation
1. Narrow band along front
2. Short duration
c. After passage
1. Wind becomes north-northwest
2. Drop in temperature—cP air mass moves in
3. Clearing skies

- D. Role of airflow aloft
  - 1. Cyclones and anticyclones
    - a. Generated by upper-level air flow
    - b. Maintained by upper-level air flow
  - 2. Cyclone
    - a. Low pressure system
    - b. Surface convergence
    - c. Outflow (divergence) aloft sustains the low pressure
  - 3. Anticyclone
    - a. High pressure system
    - b. Associated with cyclones
    - c. Surface divergence
    - d. Convergence aloft

IV. Severe weather types

- A. Thunderstorm
  - 1. Features
    - a. Cumulonimbus clouds
    - b. Heavy rainfall
    - c. Lightning
    - d. Occasional hail
  - 2. Occurrence
    - a. 2000 in progress at any one time
    - b. 100,000 per year in United States
    - c. Most frequent in
      - 1. Florida
      - 2. Eastern Gulf Coast region
  - 3. Stages of development
    - a. All thunderstorms require
      - 1. Warm air
      - 2. Moist air
      - 3. Instability (lifting)
        - a. High surface temperatures
        - b. Most common in afternoon
    - b. Require continual supply of warm air
      - 1. Each surge causes air to rise higher
      - 2. Updrafts and downdrafts form
    - c. Eventually precipitation forms
      - 1. Most active stage
      - 2. Gusty winds, lightning, hail
      - 3. Heavy precipitation
    - d. Cooling effect of precipitation marks end of thunderstorm activity
- B. Tornado
  - 1. Local storm of short duration
  - 2. Features
    - a. Violent windstorm
    - b. Rotating column of air
    - c. Extends down from cumulonimbus cloud
    - d. Low pressures inside/air rushes into
    - e. Winds approach 480 km (300 mi.) per hour
  - 3. Occurrence and development
    - a. Average of 780 each year in U.S.
    - b. Most frequent from April through June
    - c. Associated with severe thunderstorms
    - d. Exact cause for tornadoes not known
    - e. Conditions for formation of tornadoes
      - 1. Occur most often along cold front
      - 2. During the spring months
      - 3. Associated with intense thunder-storms
  - 4. Characteristics
    - a. Diameter between 150 and 600 meters
    - b. Speed across landscape about 45 km/hr
    - c. Cut about a 10 km long path
    - d. Most move toward northeast
    - e. Maximum winds about 480 km/hr
    - f. Intensity measured by Fujita intensity scale
  - 5. Tornado forecasting
    - a. Difficult to forecast because of size
    - b. Tornado watch
      - 1. To alert people
      - 2. Issued when conditions are favorable
      - 3. Covers 65,000 square kilometers
    - c. Tornado warning is issued when a funnel cloud is sighted or indicated by radar
    - d. Use of Doppler radar helps accuracy and detects motion
- C. Hurricane
  - 1. Most violent storm on Earth
  - 2. To be called a hurricane
    - a. Wind speed in excess of 119 km/hr
    - b. Rotary circulation
  - 3. Features
    - a. Tropical cyclone
      - 1. Low pressure—lowest in Western Hemisphere

2. Steep pressure gradient
3. Rapid, inward-spiraling wind

b. Parts
   1. Eyewall
      a. Near center
      b. Rising air
      c. Intense convective activity
      d. Wall of cumulonimbus clouds
      e. Greatest wind speeds
      f. Heaviest rainfall
   2. Eye
      a. At very center
      b. About 20 km diameter
      c. Precipitation ceases
      d. Wind subsides
      e. Air gradually descends and heats
      f. Warmest part of storm

c. Wind speeds reach 300 km/hr

d. Generate 50 foot waves at sea

4. Cause great damage on land
5. Known by different names
   a. Typhoon in western Pacific
   b. Cyclone in Indian Ocean
6. Frequency
   a. Most (20 per year) in North Pacific
   b. Fewer than 5 in warm North Atlantic
7. Hurricane formation and decay
   a. Form in all tropical waters except
      1. South Atlantic
      2. Eastern South Pacific
   b. Energy from condensing water vapor
   c. Develop most often in late summer when warm water temperatures provide energy and moisture
   d. Initial stage not well understood
      1. Tropical depression—winds do not exceed 61 km/hr
      2. Tropical storm—winds 61 to 119 km/hr
   e. Diminish in intensity whenever
      1. Move over cooler ocean water
      2. Move onto land
      3. Flow aloft is unfavorable
8. Destruction from a hurricane
   a. Factors that affect amount of damage
      1. Strength of storm (most important)
      2. Size and population density
      3. Shape of ocean bottom near shore
   b. Categories of damage
      1. Wind damage
      2. Storm surge
         a. Large dome of water
         b. 65 to 80 kilometers long
         c. Where eye makes landfall
      3. Inland freshwater flooding from torrential rains

Vocabulary Review

Choosing from the list of key terms, furnish the most appropriate response for the following statements.

1. A(n) ____________ is an immense body of air characterized by a similarity of temperature and moisture at any given altitude.

2. A tropical cyclonic storm having winds in excess of 119 kilometers (74 miles) per hour is called a(n) ____________.

3. A(n) ____________ is a boundary that separates different air masses, one warmer than the other and often higher in moisture content.

4. By international agreement, a(n) ____________ is a tropical cyclone with winds between 61 and 119 kilometers (38 and 74 miles) per hour.

5. A(n) ____________ is the area where an air mass acquires its characteristic properties of temperature and moisture.

6. A(n) ____________ is a type of air mass that forms over land.

Key Terms

Page numbers shown in ( ) refer to the textbook page where the term first appears.

air mass (p. 348)
air-mass weather (p. 348)
cold front (p. 351)
continental (c) air mass (p. 348)
Doppler radar (p. 362)
eye (p. 364)
eye wall (p. 364)
front (p. 349)
hurricane (p. 363)
lightning (p. 356)
maritime (m) air mass (p. 348)
middle-latitude cyclone (p. 353)
polar (P) air mass (p. 348)
source region (p. 348)
storm surge (p. 366)
thunderstorm (p. 356)
tornado (p. 357)
tornado warning (p. 361)
tornado watch (p. 361)
tropical (T) air mass (p. 348)
tropical depression (p. 365)
tropical storm (p. 365)
warm front (p. 350)

6. A(n) ____________ is a type of air mass that forms over land.

7. When the surface position of a front moves so that warm air occupies territory formerly covered by cooler air, it is called a(n) ____________.

8. In the region between southern Florida and Alaska, the primary weather producer is the ____________.

9. The ____________ is the doughnut-shaped area of intense cumulonimbus development and very strong winds that surrounds the center of a hurricane.

10. An air mass that forms in low latitudes is called a(n) ____________.

11. By international agreement, a(n) ____________ is a tropical cyclone with maximum winds that do not exceed 61 kilometers (38 miles) per hour.

12. The ____________ is a zone of scattered clouds and calm, averaging about 20 kilometers in diameter, at the center of a hurricane.

13. Fairly constant weather that may take several days to traverse an area often represents a weather situation called ____________.

14. When cold air is actively advancing into a region occupied by warmer air, the boundary is called a(n) ____________.

15. A(n) ____________ is a small, very intense cyclonic storm with exceedingly high winds, most often produced along cold fronts in conjunction with severe thunderstorms.

16. A(n) ____________ is a type of air mass that originates over water.

17. When conditions appear favorable for tornado formation, a ____________ is issued for areas covering about 65,000 square kilometers (25,000 miles).

18. The new generation of radar that can detect motion directly, and hence greatly improve tornado and severe storm warnings, is called ____________.

19. A(n) ____________ is a storm of relatively short duration produced by a cumulonimbus cloud and accompanied by strong wind gusts, heavy rain, lightning, thunder, and sometimes hail.

20. A(n) ____________ is a dome of water that sweeps across the coast near the point where the eye of the hurricane makes landfall.

21. A(n) ____________ is issued when a tornado funnel cloud has actually been sighted or is indicated by radar.

22. An air mass that originates in high latitudes is called a(n) ____________.

Comprehensive Review

1. List and describe the two criteria used to classify air masses.

   1)

   2)

2. Using Figure 14.1, list the name and describe the temperature and moisture characteristics of the air mass identified on the figure with each of the following letters.

   A:

   B:

   F:

   G:

3. Which two air masses have the greatest influence on the weather of North America, especially east of the Rocky Mountains?

4. Sketch profiles (side-views) of a typical warm front and a typical cold front and briefly describe each.

   a) Warm front:

   b) Cold front:

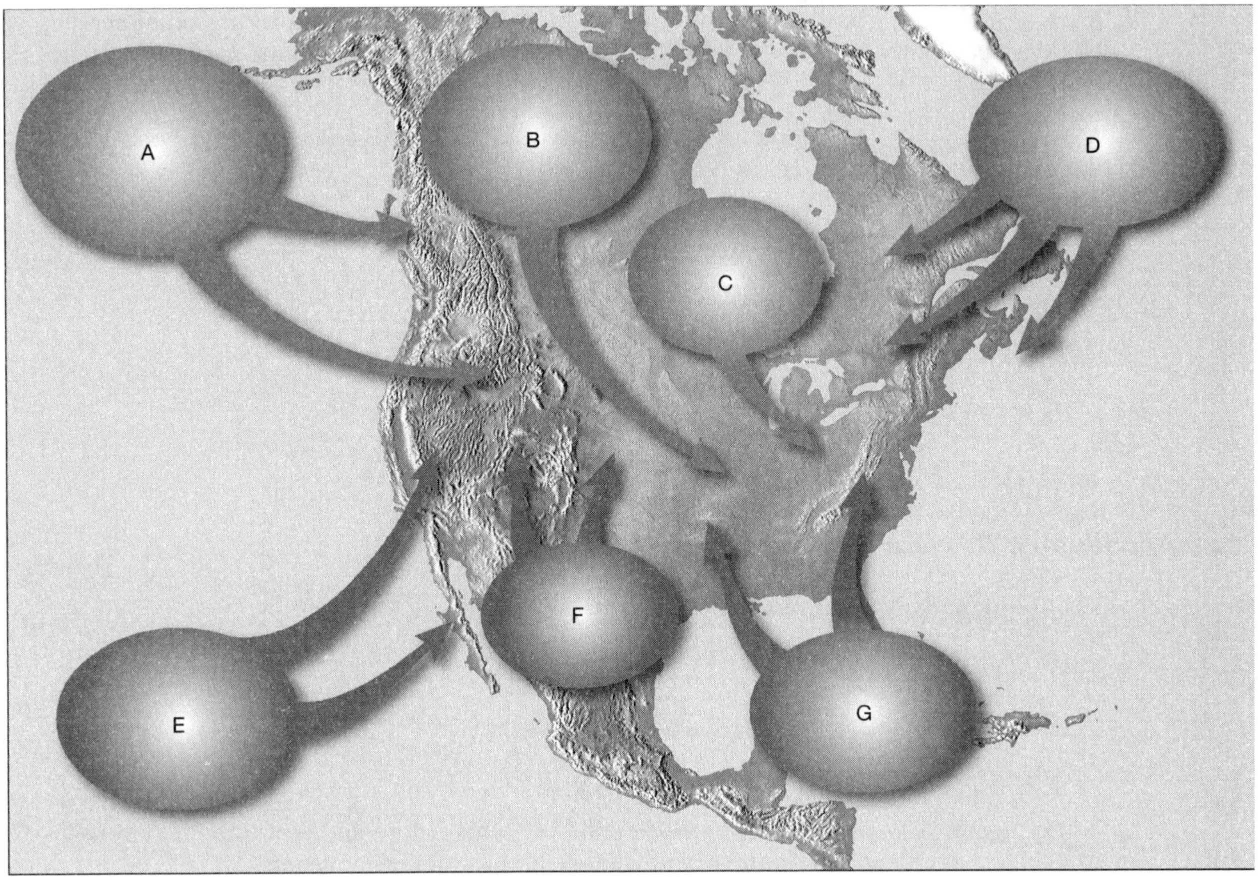

Figure 14.1

5. Using Figure 14.2, a diagram of a typical mature, middle-latitude cyclone, answer the following:

   a) Which type of front is shown at B?

   b) Which type of front is shown at D?

   c) Which air mass type would most likely be found at C?

   d) Which air mass type would most likely be found at G?

   e) The wind direction at C would most likely be from the ____________.

   f) The wind direction at G would most likely be from the ____________.

   g) As the center of the low moves to F, what would be the expected weather changes experienced by people living at C?

6. Describe the airflow at the surface and aloft for each of the following pressure systems.

   a) Cyclone:

   b) Anticyclone:

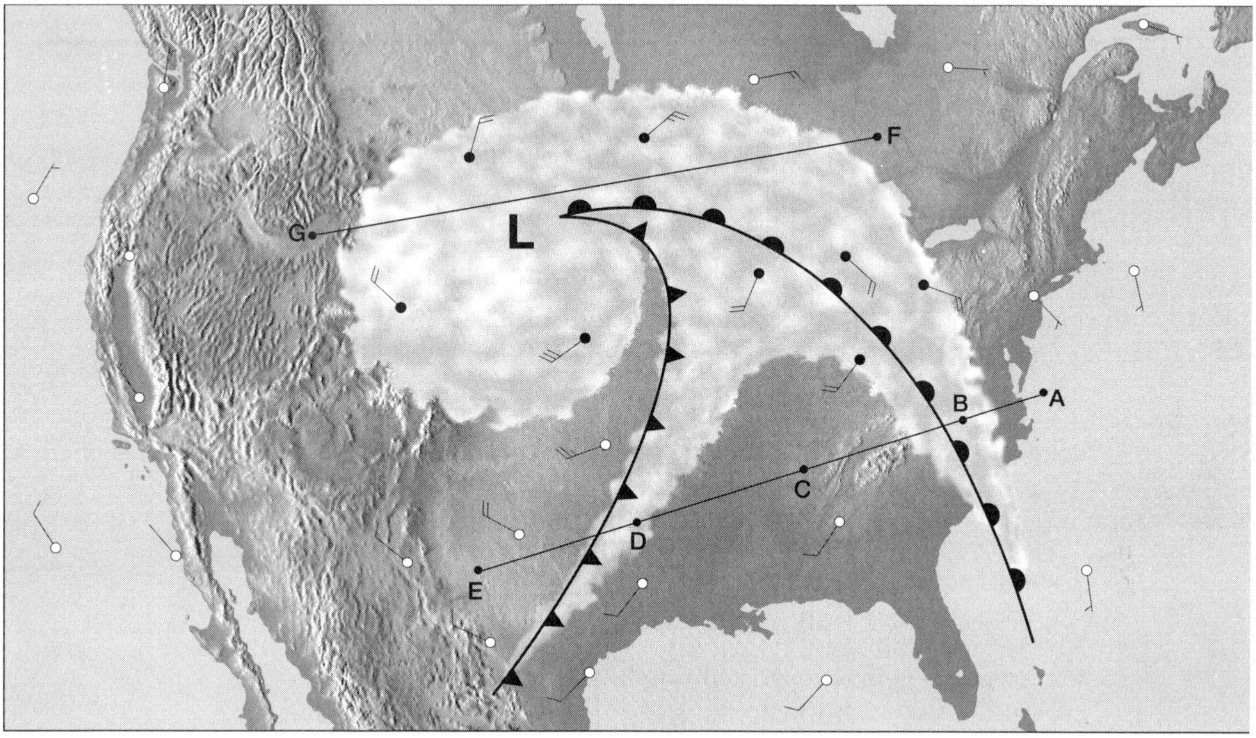

Figure 14.2

7. Compared to warm fronts, what are the two differences that largely account for the more violent nature of cold-front weather?
   1)

   2)

8. What are the atmospheric conditions associated with the formation of
   a) Thunderstorms?

   b) Tornadoes?

9. What are three conditions that can cause the intensity of hurricanes to diminish?
   1)

   2)

   3)

10. List the three categories of damage caused by hurricanes.
    1)

    2)

    3)

Practice Test

Multiple choice. Choose the best answer for the following multiple choice questions.

1. An immense body of air characterized by a similarity of temperature and moisture at any given altitude is referred to as a(n) ____________.
   a) cyclone c) anticyclone e) front
   b) air mass d) air cell

2. Hurricanes are classified according to intensity using the ____________.
   a) Richter scale c) Fujita intensity scale e) F-scale
   b) Saffir-Simpson scale d) Doppler scale

3. Along which type of front is the intensity of precipitation greater, but the duration shorter?
   a) warm front b) cold front

4. The boundary that separates different air masses is called a(n) ____________.
   a) front c) storm e) anticyclone
   b) cyclone d) contact slope

5. Which air mass is most associated with lake-effect snows?
   a) mT b) cP c) cT d) mP

6. In the United States, thunderstorms typically form within ____________ air masses.
   a) mT b) cP c) cT d) mP

7. The primary tornado warning system in use today involves both ____________.
   a) observers and satellites
   b) barometers and wind vanes
   c) observers and barometers
   d) observers and radar
   e) satellites and radar

8. Which type of front has the steepest frontal surface?
   a) warm front b) cold front

9. The area in which an air mass acquires its characteristic properties of temperature and moisture is called its ____________.
   a) area of origin c) weather site e) classification region
   b) location d) source region

10. Surface airflow in a Northern Hemisphere middle-latitude cyclone is ____________.
    a) divergent and clockwise
    b) divergent and counterclockwise
    c) convergent and clockwise
    d) convergent and counterclockwise
    e) both convergent and divergent

11. The greatest tornado frequency in the United States is during the period ____________.
    a) January through March
    b) April through June
    c) July through September
    d) October through December

12. Along a front, which air is always forced aloft?
    a) cooler, denser air
    b) the driest air
    c) the wettest air
    d) the fastest moving air
    e) warmer, less dense air

13. Following the passage of a cold front, winds often come from the ____________.
    a) south to southeast
    b) east to southeast
    c) north to northwest
    d) west to southwest
    e) south to southwest

14. Which type of air mass originates in central Canada?
    a) mT
    b) cP
    c) mP
    d) cT

15. Pressure in a middle-latitude cyclone ____________.
    a) decreases toward the center
    b) remains the same everywhere
    c) increases toward the center
    d) increases then decreases toward the center
    e) decreases then increases toward the center

16. Which one of the following is NOT an area where maritime tropical air masses that affect North America originate?
    a) Gulf of Mexico
    b) Hudson Bay
    c) Caribbean Sea
    d) Atlantic Ocean
    e) Pacific Ocean

17. The first sign of the approach of a warm front is the appearance of ____________ clouds overhead.
    a) stratus
    b) cirrus
    c) cumulus
    d) cumulonimbus
    e) nimbostratus

18. The weather behind a cold front is dominated by a ____________.
    a) subsiding, relatively cold air mass
    b) rising, relatively warm air mass
    c) mixed air mass
    d) rising, relatively cold air mass
    e) subsiding, relatively warm air mass

19. An mT air mass would be best described as ____________.
    a) cold and dry
    b) warm and dry
    c) cold and wet
    d) warm and wet
    e) hot and dry

20. Hurricanes form in tropical waters between the latitudes of ____________.
    a) 0 and 5 degrees
    b) 5 and 20 degrees
    c) 20 and 30 degrees
    d) 30 and 40 degrees

21. The source of the heat necessary to maintain the upward development of a thunderstorm is ____________.
    a) surface heating
    b) friction
    c) latent heat
    d) electrical heat
    e) compressional heat

22. Which air mass is most associated with a "northeaster" in New England?
    a) mT b) cP c) mP d) cT

23. Following the passage of a warm front, winds often come from the ____________.
    a) north to northeast c) north to northwest e) south to southwest
    b) east to southeast d) west to northwest

24. Hurricanes develop most often in the ____________.
    a) early winter c) early summer e) early spring
    b) late winter d) late summer

25. The center of a tornado is best characterized by its ____________.
    a) very high pressure c) large calm area e) subsidence
    b) eye wall d) very low pressure

True/false. For the following true/false questions, if a statement is not completely true, mark it false. For each false statement, change the ***italicized*** *word to correct the statement.*

1. ___ A *front* usually marks a change in weather.
2. ___ Air masses are typically *five* thousand miles or more across.
3. ___ Cyclones are typically the bearers of *stormy* weather.
4. ___ A cP air mass originates over land and is likely to be cold and *dry.*
5. ___ Cold fronts advance *less* rapidly than warm fronts.
6. ___ Guided by the westerlies aloft, middle-latitude cyclones generally move *westward* across the United States.
7. ___ *Continental* tropical air is the source of much, if not most, of the precipitation received in the eastern two-thirds of the United States.
8. ___ Most severe weather occurs along *cold* fronts.
9. ___ Doppler radar has the ability to detect *motion* directly.
10. ___ On a weather map, the surface position of a *warm* front is shown by a line with semicircles extending into the cooler air.
11. ___ After the passage of a cold front, air pressure will most likely *rise.*
12. ___ Most severe thunderstorms in the mid-latitudes form along or ahead of *warm* fronts.
13. ___ On the average, the slopes of cold fronts are about *half* as steep as warm fronts.
14. ___ In a middle-latitude cyclone, *convergence* and forceful lifting initiate cloud development.
15. ___ Cyclones and anticyclones are typically found *adjacent* to one another.
16. ___ More tornadoes are generated in the *western* United States than any other area of the country.

17. ___ In the western Pacific, hurricanes are called *typhoons.*

18. ___ A tornado *warning* is issued when a funnel cloud has actually been sighted or indicated by radar.

19. ___ In a middle-latitude cyclone, *divergence* is occurring in the air aloft.

20. ___ A tornado with an F4 rating on the Fujita intensity scale produces *devastating* damage.

21. ___ The lowest pressures ever recorded in the Western Hemisphere are associated with *hurricanes.*

22. ___ Airflow aloft plays an *important* role in maintaining cyclonic and anticyclonic circulation.

Written questions

1. Describe the temperature and moisture characteristics of 1) a continental polar (cP) air mass and 2) a maritime tropical (mT) air mass.

2. Describe the weather conditions that an observer would experience as a middle-latitude cyclone passes with its center to the north.

3. What is the difference between a tornado watch and a tornado warning?

15 The Nature of the Solar System

The Nature of the Solar System opens by examining the observations and contributions made by early civilizations and the Greek philosophers, including Ptolemy. An in-depth examination of the birth of modern astronomy centers on the contributions of Nicolaus Copernicus, Tycho Brahe, Johannes Kepler, Galileo Galilei, and Sir Isaac Newton. An investigation of the differences between the terrestrial and Jovian planets is followed by a discussion of the origin of the solar system. Included in the hapter is a detailed study of the moon's physical characteristics. An inventory of the planets (excludng Earth) presents the prominent features and peculiarities of each planet. The chapter closes with a discussion of the minor members of the solar system—asteroids, comets, and meteoroids.

Learning Objectives

After reading, studying, and discussing this chapter, you should be able to:

- Describe the geocentric theory of the universe held by many early Greeks.
- List the contributions to modern astronomy of Nicolaus Copernicus, Tycho Brahe, Johannes Kepler, Galileo Galilei, and Sir Isaac Newton.
- Describe the general characteristics of the two groups of planets in the solar system.
- Discuss the origin of the solar system.
- Describe the major features of the lunar surface.
- List the distinguishing features of each planet in the solar system.
- List and describe the minor members of the solar system.

Chapter Review

- Early Greeks held the *geocentric* ("Earth-centered") view of the universe, believing that Earth was a sphere that stayed motionless at the center of the universe. Orbiting Earth were the moon, sun, and the known planets—Mercury, Venus, Mars, Jupiter, and Saturn. To the early Greeks, the stars traveled daily around Earth on a transparent, hollow sphere called the *celestial sphere.* In A.D. 141, *Claudius Ptolemy* presented the geocentric outlook of the Greeks in its finest form which became known as the *Ptolemaic system.*

- Modern astronomy evolved through the work of many dedicated individuals during the 1500s and 1600s. *Nicolaus Copernicus* (1473-1543) reconstructed the solar system with the sun at the center and the planets orbiting around it, but erroneously continued to use circles to represent the orbits of planets. *Tycho Brahe's* (1546-1601) observations were far more precise than any made previously and are his legacy to astronomy. *Johannes Kepler* (1571-1630) ushered in the new astronomy with his three laws of planetary motion. After constructing his own telescope, *Galileo Galilei* (1564-1642) made many important discoveries that supported the Copernican view of a sun-centered solar system. *Sir Isaac Newton* (1643-1727) developed laws of motion and proved that the force of gravity, combined with the tendency of an object to move in a straight line, results in elliptical orbits for planets.

• The planets can be arranged into two groups: the *terrestrial* (Earthlike) *planets* (Mercury, Venus, Earth, and Mars) and the *Jovian* (Jupiterlike) planets (Jupiter, Saturn, Uranus, and Neptune). Pluto is not included in either group. When compared to the Jovian planets, the terrestrial planets are smaller, more dense, contain proportionally more rocky material, and have slower rates of rotation.

• The *nebular hypothesis* describes the formation of the solar system. The planets and sun began forming about 5 billion years ago from a large cloud of dust and gases called a *nebula*. As the nebular cloud contracted, it began to rotate and assume a disk shape. Material that was gravitationally pulled toward the center became the *protosun*. Within the rotating disk, small centers, called *protoplanets*, swept up more and more of the nebular debris. Due to their high temperatures and weak gravitational fields, the inner planets (Mercury, Venus, Earth, and Mars) were unable to accumulate many of the lighter components (hydrogen, ammonia, methane, and water) of the nebula. However, because of the very cold temperatures existing far from the sun, the fragments from which the Jovian planets formed contained a high percentage of ices—water, carbon dioxide, ammonia, and methane.

• The lunar surface exhibits several types of features. Most *craters* were produced by the impact of rapidly moving debris (meteoroids). Bright, densely cratered *highlands* make up most of the lunar surface. Dark, fairly smooth lowlands are called *maria*. Maria basins are enormous impact craters that have been flooded with layer upon layer of very fluid basaltic lava. All lunar terrains are mantled with a soil-like layer of gray, unconsolidated debris, called *lunar regolith*, which has been derived from a few billion years of meteoric bombardment.

• *Mercury* is a small, dense planet that has no atmosphere and exhibits the greatest temperature extremes of any planet. *Venus*, the brightest planet in the sky, has a thick, heavy atmosphere composed primarily of carbon dioxide, a surface of relatively subdued plains and inactive volcanic features, a surface atmospheric pressure ninety times that of Earth's, and surface temperatures of 475°C (900°F). *Mars*, the red planet, has a carbon dioxide atmosphere only 1 percent as dense as Earth's, extensive dust storms, numerous inactive volcanoes, many large canyons, and several valleys of debatable origin exhibiting drainage patterns similar to stream valleys on Earth. *Jupiter*, the largest planet, rotates rapidly, has a banded appearance, a Great Red Spot that varies in size, a ring system, and at least sixteen moons (one of the moons, Io, is a volcanically active body). *Saturn* is best known for its system of rings. It also has a dynamic atmosphere with winds up to 930 miles per hour and "storms" similar to Jupiter's Great Red Spot. *Uranus* and *Neptune* are often called "the twins" because of similar structure and composition. A unique feature of Uranus is the fact that it rotates "on its side." Neptune has white, cirruslike clouds above its main cloud deck and an Earth-sized Great Dark Spot, assumed to be a large rotating storm similar to Jupiter's Great Red Spot. *Pluto*, a small frozen world with one moon, may have once been a satellite of Neptune. Pluto's noticeably elongated orbit causes it to occasionally travel inside the orbit of Neptune, but with no chance of collision.

• The minor members of the solar system include the *asteroids*, *comets*, and *meteoroids*. No conclusive evidence has been found to explain the origin of the asteroids. Comets are made of frozen gases with small pieces of rocky and metallic material. Many travel in very elongated orbits that carry them beyond Pluto. Meteoroids, small solid particles that travel through interplanetary space, become meteors when they enter Earth's atmosphere and vaporize with a flash of light. *Meteor showers* appear to occur when Earth encounters a swarm of meteoroids, probably lost by a comet. *Meteorites* are the remains of meteoroids found on Earth. The *three types of meteorites* are 1) *iron*, 2) *stony*, and 3) *stony-iron*.

Chapter Outline

I. Early history of astronomy
  A. Ancient Greeks
    1. Used philosophical arguments
    2. Some observational data
    3. Held geocentric view
      a. "Earth-centered"
        1. Earth center of universe
        2. Stars on celestial sphere traveled daily around Earth
        3. Seven wanderers ("planetai")
          a. Changed position in sky
          b. Included
            1. Sun
            2. Moon
            3. Mercury through Saturn (excluding Earth)
          c. Planets exhibit apparent westward drift
            1. Called retrograde motion
            2. From planet's and Earth's motions
        4. Ptolemaic system
          a. A.D. 141
          b. Geocentric model
          c. To explain retrograde motion he had planets moving in
            1. Deferents (large circles), and
            2. Epicycles (small circles)
  B. Birth of modern astronomy
    1. 1500s and 1600s
    2. Five noted scientists
      a. Nicolaus Copernicus (1473-1543)
        1. Model put sun at center
        2. Used circular orbits for planets
        3. Ushered out old astronomy
      b. Tycho Brahe (1546-1601)
        1. Precise observer
        2. Tried to find stellar parallax—shift in a star's position due to revolution of Earth
        3. Did not believe in Copernican system—unable to observe stellar parallax
      c. Johannes Kepler (1571-1630)
        1. Ushered in new astronomy
        2. Planets revolve around sun
        3. Three laws of planetary motion
          a. Orbits of planets are elliptical
          b. Planets revolve at varying speed
          c. Proportional relation between a planet's orbital period and its distance to the sun (unit of distance: astronomical unit)
      d. Galileo Galilei (1564-1642)
        1. Supported Copernican theory
        2. Used experimental data
        3. Constructed astronomical telescope
        4. Discoveries using telescope
          a. Four large moons of Jupiter
          b. Planets appeared as disks
          c. Phases of Venus
          d. Features on moon
          e. Sunspots
        5. Tried and convicted during the Inquisition
      e. Sir Isaac Newton (1643-1727)
        1. Law of universal gravitation
        2. Explained planetary motion

II. Solar system
  A. Overview
    1. Solar system includes
      a. Sun
      b. Nine planets and their satellites
      c. Asteroids
      d. Comets
      e. Meteoroids
    2. Planet's orbits lie in orbital plane
    3. Solar system composed of
      a. Gases—hydrogen and helium
      b. Rock
        1. Silicate minerals
        2. Metallic iron
      c. Ices
        1. Ammonia
        2. Methane
        3. Carbon dioxide
        4. Water
    4. Two groups of planets
      a. Terrestrial (Earthlike) planets
        1. Mercury through Mars
        2. Small, dense, rocky
        3. Low escape velocities

- b. Jovian (Jupiterlike) planets
  - 1. Jupiter through Neptune
  - 2. Large, low density, gaseous
  - 3. High escape velocities
- c. Pluto not included in either group

B. Origin
- 1. Nebular hypothesis
  - a. Planets formed at same time 1.5 billion years ago
  - b. Large cloud of dust and gas called nebula
    - 1. 80% hydrogen
    - 2. 15% helium
    - 3. All other heavier elements
  - c. Nebula contracted
  - d. Nebula rotates
  - e. Protosun forms
  - f. Protoplanets form

C. Moon
- 1. Diameter of 2150 miles
- 2. Density less than Earth's
- 3. Small iron core
- 4. Lunar surface features
  - a. Crater
    - 1. Most produced by impact from meteoroids
    - 2. Impact produces
      - a. Ejecta
      - b. Occasional rays
  - b. Two types of terrains
    - 1. Highland
      - a. Bright regions
      - b. Densely cratered
      - c. Most of lunar surface
    - 2. Mare
      - a. Dark region
      - b. Fairly smooth lowland
      - c. Large impact crater filled with fluid basaltic lava
  - c. Lunar regolith
    - 1. Covers all lunar terrains
    - 2. Gray, unconsolidated debris
    - 3. Soil-like layer
    - 4. Produced by meteoric bombardment

D. Planets
- 1. Mercury
  - a. Innermost planet
  - b. Smallest
  - c. No atmosphere
  - d. Cratered highlands
  - e. Vast, smooth terrains
  - f. Very dense
  - g. Rotates slowly
    - 1. Cold nights (-280°F)
    - 2. Hot days (800°F)
- 2. Venus
  - a. Second to moon in brilliance
  - b. Similar to Earth in
    - 1. Size
    - 2. Density
    - 3. Location
  - c. Thick clouds
    - 1. Impenetrable by visible light
    - 2. 97% carbon dioxide
    - 3. Surface atmospheric pressure 90 times that of Earth's
  - d. Surface
    - 1. Mapped by radar
    - 2. Features
      - a. 80% of surface is subdued plains mantled by volcanic flows
      - b. Low density of impact craters
      - c. Tectonic deformation
      - d. Volcanic structures
- 3. Mars
  - a. Atmosphere
    - 1. 1% as dense as Earth's
    - 2. Primarily carbon dioxide
    - 3. Cold polar temperatures (-193°F)
    - 4. Polar caps of solidified carbon dioxide
    - 5. Dust storms
      - a. Winds up to 170 miles per hour
      - b. Surface
        - 1. Large volcanoes—largest is Mons Olympus
        - 2. Canyons
          - a. Some larger than Grand Canyon
          - b. Largest—Valles Marineras
            - 1. Almost 5000 km long
            - 2. Formed from huge faults
      - c. Stream drainage patterns
        - 1. In some valleys
        - 2. No surface water on planet

            3. Possible origins
                a. Past rainfall
                b. Seepage of groundwater
                c. Surface material collapse
    4. Jupiter
        a. Largest planet
        b. Massive
        c. Rapid rotation
            1. Slightly less than 10 hours
            2. Slightly bulged equatorial region
        d. Banded appearance
        e. Great Red Spot
            1. In planet's southern hemisphere
            2. 11,000 km by 22,000 km
            3. Counterclockwise rotation
            4. Perhaps a large cyclonic storm
        f. Ring system
        g. Moons
            1. At least 16 moons
            2. Four largest moons
                a. Discovered by Galileo
                b. Called Galilean moons
                c. Each has its own character
                    1. Callisto
                        a. Densely cratered
                        b. Frozen water ice surface
                    2. Europa
                        a. Smallest Jupiter moon
                        b. Icy surface
                        c. Linear surface features
                    3. Ganymede
                        a. Largest in solar system
                        b. Diverse terrains
                        c. Parallel grooves show past tectonic history
                    4. Io
                        a. Innermost Jupiter moon
                        b. Volcanically active—heat from tidal energy
                        c. Sulfurous
    5. Saturn
        a. Similar to Jupiter in its
            1. Atmosphere
            2. Composition
            3. Internal structure
        b. Rings
            1. Most prominent feature
            2. Complex structure
            3. Origin—particles never became satellite
            4. A few hundred meters thick
            5. Lateral extent exceeds 200,000 km
            6. Classically called
                a. A ring
                    1. Outermost ring
                    2. Cassini division
                        a. Large gap (5000 km)
                        b. Separates A and B rings
                b. B ring—brightest ring
                c. C ring
        c. Dynamic atmosphere
        d. Over 20 moons
            1. Titan
                a. Second largest in solar system—larger than Mercury—has substantial atmosphere
                    1. 80% nitrogen
                c. May have polar ice caps
        e. Large cyclonic "storms"
    6. Uranus
        a. Uranus and Neptune considered twins
        b. Rotates "on its side"—axis inclined almost 90°
        c. Rings
        d. Large moons have varied terrains
    7. Neptune
        a. Dynamic atmosphere
            1. Great Dark Spot
            2. White clouds above main clouds
        b. Eight satellites
            1. Triton
                a. Largest Neptune moon
                b. Highly inclined retrograde orbit
            2. Other moons have a variety of surfaces
        c. Lowest solar system temperature—1.38 K (-391°F)
        d. Thin atmosphere
        e. Volcaniclike activity
    8. Pluto
        a. Not visible with unaided eye
        b. Discovered in 1930
        c. Moon (Charon) discovered in 1978
        d. Average temperature is -210°C
        e. May have been moon of Neptune
E. Minor members of the solar system

1. Asteroids
   a. Most between Mars and Jupiter
   b. Small bodies
      1. Largest (Ceres)—about 620 miles in diameter
   c. Some have very eccentric orbits and impact the moon and Earth
   d. Irregular shapes
   e. Origin is uncertain
2. Comets
   a. Large, "dirty snowballs"
      1. Frozen gases
      2. Rocky and metallic materials
   b. Frozen gases vaporize when near sun
      1. Glowing head (coma) produced
      2. Some may develop a tail that points away from the sun due to
         a.Radiation pressure
         b.Solar wind
   c. Origin
      1.Not well known
      2.Form at great distance from sun
   d. Comet Shoemaker-Levy impacted Jupiter in July 1994 and caused impact flashes, debris plumes, and dark zones

e. Meteor shower
   1. Earth passes through a comet's orbit
   2. Particles vaporize in atmosphere

f. Meteoroids
   1. Called meteors when they enter Earth's atmosphere
   2. Referred to as meteorites when they are found on Earth
   3. Types based on composition
      a. Irons
         1. Mostly iron
         2. 5-20% nickel
      b. Stony—mostly silicate minerals
      c. Stony-irons—mixtures
      d. Carbonaceous chondrites
         1. Simple amino acids
         2. Other organic material
   4. Give an idea as to Earth's core of iron and nickel composition
   5. Give an idea as to solar system age which exceeds 4.5 billion years

Vocabulary Review

Choosing from the list of key terms, furnish the most appropriate response for the following statements.

1. The apparent westward drift of the planets with respect to the stars is called ____________.

2. A(n) ____________ is a planet that has physical characteristics similar to those of Earth.

3. A(n) ____________ is a small, planetlike body whose orbit lies mainly between Mars and Jupiter.

4. The dark regions on the moon that resemble "seas" on Earth are ____________.

5. Early Greeks held the ____________ view of the universe, believing that Earth was a sphere that stayed motionless at its center.

6. The physical characteristics of a(n) ____________ are like those of Jupiter.

7. The average distance from Earth to the sun, about 150 million kilometers (93 million miles), is a unit of distance called the ____________.

8. ____________ is the soil-like layer of gray, unconsolidated debris that mantles all lunar terrains.

9. A(n) ____________ is a luminous phenomenon observed when a small solid particle enters Earth's atmosphere and burns up.

10. The initial speed that an object needs before it can leave a planet and go into space is known as the ___________.

11. A solid fragment from space that strikes Earth's surface and contains mostly silicate minerals (with inclusions of other minerals) is called a(n) ___________.

12. A small body made of frozen gases and small pieces of rocky and metallic materials which generally revolves about the sun in an elongated orbit is a(n) ___________.

13. The ___________ was an imaginary, transparent, hollow sphere on which the ancients believed the stars traveled daily around Earth.

14. The remains of a meteoroid, when found on Earth, is referred to as a(n) ___________.

15. A spectacular display of numerous meteor sightings that occurs when Earth encounters a swarm of meteoroids is called a(n) ___________.

Key Terms

Page numbers shown in ( ) refer to the textbook page where the term first appears.

asteroid (p. 401)
astronomical unit (AU) (p. 379)
Cassini division (p. 397)
celestial sphere (p. 374)
comet (p. 402)
escape velocity (p. 384)
geocentric (p. 374)
iron meteorite (p. 405)
Jovian planet (p. 382)
lunar regolith (p. 388)
maria (p. 386)
meteor (p. 404)
meteorite (p. 405)
meteor shower (p. 404)
nebular hypothesis (p. 384)
Ptolemaic system (p. 375)
retrograde motion (p. 375)
stony-iron meteorite (p. 405)
stony meteorite (p. 405)
terrestrial planet (p. 382)

16. The ___________ is an Earth-centered model of the universe proposed in A.D. 141 which uses epicycles and deferents to describe a planet's motion.

17. The large gap that separates Saturn's A and B rings, which can be viewed from Earth, is called the ___________.

18. A meteorite that contains mostly iron with 5-20 percent nickel is called a(n) ___________.

19. The ___________ suggests that all bodies of the solar system formed from an enormous cloud of gases and dust.

20. A meteorite that is a mixture of iron, nickel, and silicate minerals is called a(n) ___________.

Comprehensive Review

1. Describe the geocentric model of the universe held by the early Greeks.

2. What is retrograde motion? How did the geocentric model of Ptolemy explain retrograde motion?

3. Why did early Greeks reject the idea of a rotating Earth?

4. For each of the following statements, list the name of the scientist credited with the contribution.
   a) First to use the telescope for astronomy:

   b) Proposed circular orbits of the planets around the sun in 1500s:

   c) Systematically measured the locations of heavenly bodies:

   d) Proposed laws of planetary motion:

   e) Monumental work, *De Revolutionibus*:

   f) Proposed elliptical orbits for planets around the sun:

   g) Discovered four satellites, or moons, orbiting Jupiter:

   h) Wrote *Dialogue of the Great World Systems*:

   i) Law of universal gravitation:

5. Describe stellar parallax.

6. List each of Kepler's three laws of planetary motion.
   1)

   2)

   3)

7. List four of Galileo's astronomical discoveries.
    1)

    2)

    3)

    4)

8. Describe the following two groups of planets and list the planets that are included in each group.
    a) Terrestrial planets:

    b) Jovian planets:

9. For each of the following characteristics, write a brief statement that compares the terrestrial planets to the Jovian planets.
    a) Size:

    b) Density:

    c) Period of rotation:

    d) Mass:

    e) Number of known satellites:

    f) Period of revolution:

    g) General composition:

10. What are the gas, rock, and ice substances that compose the majority of the solar system?
    a) Gases:

    b) Rocks:

    c) Ices:

11. Briefly describe the origin of the solar system according to the nebular hypothesis. What were the two gases that composed the majority of the nebula that formed the solar system?

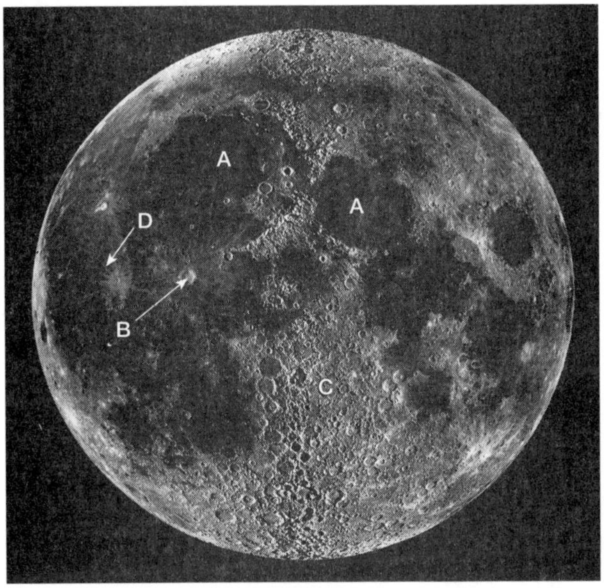

Figure 15.1

12. Using the photograph of the moon, Figure 15.1, select the letter that identifies each of the following lunar features.
    a) Mare: ____
    b) Crater: ____
    c) Ray(s): ____
    d) Highland: ____
    e) Oldest feature: ____

13. Why would a 100-pound person weigh only 17 pounds on the moon?

14. Describe the origin of lunar maria.

15. List the name of the planet described by each of the following statements.
    a) Red planet:

    b) Thick clouds, dense carbon dioxide atmosphere, volcanic surface:

    c) Greatest diameter:

    d) Has the only satellite (Titan) with a substantial atmosphere:

    e) Hot surface temperature and no atmosphere:

    f) Cassini division:

    g) Second only to the moon in brilliance in the night sky:

    h) Axis of rotation lies near the plane of its orbit:

    i) Great Red Spot:

    j) Moon (Triton) has the lowest surface temperature in the solar system:

    k) Most prominent system of rings:

    l) Carbon dioxide atmosphere 1 percent as dense as Earth's, volcanic surface:

    m) Moons include Io, Callisto, Ganymede, and Europa:

    n) "Earth's twin":

    o) Great Dark Spot:

    p) Revolves quickly but rotates slowly:

    q) Longest period of revolution:

    r) Explored by *Sojourner* in 1997:

16. What are the two hypotheses for the origin of the asteroids?
    1)

    2)

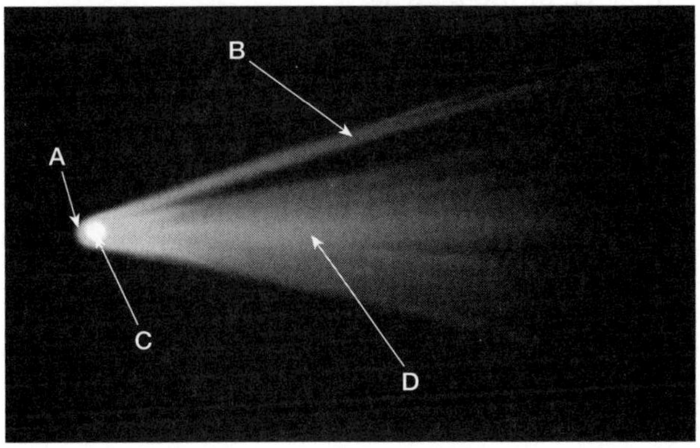

Figure 15.2

17. Using Figure 15.2, list the letter that identifies each of the following comet parts.
    a) Ionized gas tail: ____
    b) Coma: ____
    c) Nucleus: ____
    d) Dust tail: ____

18. What are the two forces that contribute to the formation of a comet's tail?
    1)

    2)

19. What is the difference between a meteor and a meteorite?

20. List and briefly describe the three types of meteorites based on composition.

    1)

    2)

    3)

Practice Test

Multiple choice. Choose the best answer for the following multiple choice questions.

1. Who was the ancient Greek that developed a geocentric model of the universe explaining the observable motions of the planets?
   a) Aristotle
   b) Ptolemy
   c) Copernicus
   d) Kepler
   e) Newton

2. Which one of the following is NOT a terrestrial planet?
   a) Mercury
   b) Earth
   c) Mars
   d) Venus
   e) Jupiter

3. The fact that Mercury is very dense implies that it contains a(n) ____________.
   a) asthenosphere
   b) liquid interior
   c) magma reservoir
   d) iron core
   e) highly volatile surface

4. One astronomical unit averages about ____________.
   a) 93 million miles
   b) 210 million kilometers
   c) 39 million miles
   d) 93 million kilometers
   e) 150 million miles

5. The large dark regions on the moon are ____________.
   a) highlands
   b) craters
   c) mountains
   d) maria
   e) rays

6. Ancient astronomers believed that ____________.
   a) Earth revolved around the celestial sphere
   b) the sun was the center of the universe
   c) the sun was on the celestial sphere
   d) Earth was a "wanderer"
   e) Earth was the center of the universe

7. Which planet has a mass greater than the combined mass of all the remaining planets, satellites, and asteroids?
   a) Venus
   b) Mars
   c) Jupiter
   d) Uranus
   e) Saturn

8. A very large interstellar cloud of dust and gases is called a(n) ____________.
   a) nebula
   b) epicycle
   c) protoplanet
   d) gas core
   e) protosun

9. The apparent westward movement of a planet against the background of stars is called ____________.
   a) retrograde motion
   b) reverse motion
   c) westward drift
   d) Ptolemaic motion
   e) deferent motion

10. Which one of the following is NOT a material commonly found in lunar regolith?
    a) glass beads
    b) clay
    c) lunar dust
    d) igneous rocks

11. During the formation of the solar system, the inner planets were unable to accumulate much hydrogen, ammonia, methane, and water because of their ____________.
    a) high temperatures and weak magnetism
    b) high temperatures and weak gravities
    c) high magnetism and slow velocities
    d) low temperatures and high velocities
    e) low temperatures and strong gravities

12. Which one of the following planets was unknown to the ancient Greeks?
    a) Earth c) Uranus e) Venus
    b) Mars d) Mercury

13. Which one of Jupiter's moons is volcanically active?
    a) Callisto b) Europa c) Ganymede d) Io

14. Using Tycho Brahe's data, which scientist proposed three laws of planetary motion?
    a) Newton c) Kepler e) Ptolemy
    b) Galileo d) Copernicus

15. Which chemical is responsible for the greenish-blue color of Uranus and Neptune?
    a) hydrogen sulfide c) ammonia e) carbon dioxide
    b) methane d) hydrogen chloride

16. Which planet has a density less than that of water?
    a) Mercury c) Jupiter e) Neptune
    b) Venus d) Saturn

17. Which planet might best be described as a large, dirty iceball?
    a) Mercury c) Mars e) Pluto
    b) Venus d) Saturn

18. The first modern astronomer to propose a sun-centered solar system was ____________.
    a) Galileo c) Ptolemy e) Brahe
    b) Newton d) Copernicus

19. Although this planet is shrouded in thick clouds, radar mapping has revealed a varied topography.
    a) Mercury c) Mars e) Uranus
    b) Venus d) Jupiter

20. Which scientist discovered that Venus has phases, just like the moon?
    a) Ptolemy c) Galileo e) Newton
    b) Copernicus d) Kepler

21. Which moon is the only satellite in the solar system with a substantial atmosphere?
    a) Io c) Miranda e) Europa
    b) Triton d) Titan

22. The apparent shift in the position of a star caused by Earth's motion is called ____________.
    a) stellar shift c) an epicycle e) revolution
    b) retrograde motion d) stellar parallax

23. The moon's density is comparable to that of Earth's ____________.
    a) core
    b) atmosphere
    c) crustal rocks
    d) asthenosphere
    e) lower mantle

24. Which planet's atmosphere contains the Great Dark Spot?
    a) Venus
    b) Jupiter
    c) Saturn
    d) Uranus
    e) Neptune

25. Which feature(s) on Mars have raised the question about the possibility of liquid water on the planet?
    a) mountain ranges with faults
    b) impact craters with sharp rims
    c) volcanic cones with craters
    d) deep, long canyons
    e) valleys with tributaries

26. The Jovian planets contain a large percentage of the gases ____________.
    a) nitrogen and argon
    b) hydrogen and oxygen
    c) oxygen and nitrogen
    d) helium and oxygen
    e) hydrogen and helium

27. Which one of the following is NOT included in the solar system?
    a) comets
    b) galaxies
    c) sun
    d) planets
    e) asteroids

28. Second only to the moon in brilliance in the night sky is ____________.
    a) Mercury
    b) Venus
    c) Mars
    d) Jupiter
    e) Saturn

29. According to the ancients, the stars traveled around Earth on the transparent, hollow, ________.
    a) deferent
    b) celestial sphere
    c) equatorial sphere
    d) solar orb
    e) stellar globe

30. Although the atmosphere of this planet is very thin, extensive dust storms do occur with wind speeds in excess of 150 miles per hour.
    a) Mercury
    b) Venus
    c) Mars
    d) Saturn
    e) Uranus

31. Which planet's axis lies only 8 degrees from the plane of its orbit?
    a) Mercury
    b) Mars
    c) Jupiter
    d) Saturn
    e) Uranus

32. Which planet completes one rotation in slightly less than 10 hours?
    a) Mercury
    b) Venus
    c) Mars
    d) Jupiter
    e) Saturn

33. Relatively young lunar craters often exhibit ____________.
    a) rays
    b) maria
    c) eroded rims
    d) highlands
    e) active volcanoes

34. Which one of the following planets does NOT have rings?
    a) Mars
    b) Jupiter
    c) Saturn
    d) Uranus
    e) Neptune

35. Through the telescope, this planet appears as a reddish ball interrupted by some permanent dark regions that change intensity.
    a) Mercury c) Mars e) Uranus
    b) Venus d) Saturn

36. Most asteroids lie between the orbits of ___________.
    a) Mercury and Venus c) Earth and Mars e) Jupiter and Saturn
    b) Venus and Earth d) Mars and Jupiter

37. Most meteor showers are associated with the orbits of ___________.
    a) satellites c) planets e) meteorites
    b) comets d) asteroids

True/false. For the following true/false questions, if a statement is not completely true, mark it false. For each false statement, change the ***italicized*** *word to correct the statement.*

1. ___ The orbits of the planets are *circular.*
2. ___ The most prominent feature of *Pluto* is its ring system.
3. ___ Compared to other planet-satellite systems, Earth's moon is unusually *large.*
4. ___ The farther away a star is, the *greater* will be its parallax.
5. ___ *Earth* has the greatest temperature extremes of any planet.
6. ___ The force of gravity is *inversely* proportional to the square of the distance between two objects.
7. ___ The Jovian planets have *high* densities and large masses.
8. ___ Bombardment of the moon by *micrometeorites* gradually smooths the landscape.
9. ___ A *geocentric* model holds that Earth is the center of the universe.
10. ___ The Martian atmosphere is *fifty* percent as dense as that of Earth.
11. ___ Most of the lunar surface is made up of *maria.*
12. ___ The orbital periods of the planets and their *distances* to the sun are proportional.
13. ___ The idea that all bodies in the solar system formed from an enormous cloud is referred to as the *protosun* hypothesis.
14. ___ The most striking feature on the face of *Mars* is the Great Red Spot.
15. ___ The escape velocity for Earth is *seven* miles per second.
16. ___ The formation of rings is believed be related to the *gravitational* force of a planet.

17. ___ The *surface* of Venus reaches temperatures of 475°C (900°F).

18. ___ The *retrograde* motion of Triton suggests that it formed independently of Neptune.

19. ___ In 1978, the moon Charon was discovered orbiting *Pluto.*

20. ___ The *Voyager 2* spacecraft has surveyed the greatest number of planets, including Neptune in 1989.

Written questions

1. Compare the physical characteristics of the terrestrial planets to those of the Jovian planets.

2. Briefly describe origin of the solar system according to the nebular hypothesis.

3. Distinguish between a meteoroid, meteor, and meteorite.

16

Beyond the Solar System

Beyond the Solar System begins with an examination of the intrinsic properties of stars—distance, brightness, color, and temperature. Binary star systems, stellar mass, and the Hertzsprung-Russell diagram are discussed in detail. Also investigated are the various types of nebulae. Stellar evolution, from birth through protostar, main-sequence, red giant, burnout and death, is presented. Following stellar evolution are descriptions of the various stellar remnants—dwarf stars, neutron stars, and black holes. The chapter continues with a detailed discussion of the Milky Way galaxy and a general description of types of galaxies, galactic clusters, and red shifts. The chapter closes with a commentary on the Big Bang theory.

Learning Objectives

After reading, studying, and discussing this chapter, you should be able to:

- Discuss the principle of parallax and explain how it is used to measure the distance to a star.
- List and describe the major intrinsic properties of stars.
- Describe the different types of nebulae.
- Describe the most plausible model for stellar evolution and list the stages in the life cycle of a star.
- Describe the possible final states that a star may assume after it consumes its nuclear fuel and collapses.
- List and describe the major types of galaxies.
- Describe the Big Bang theory of the origin of the universe.

Chapter Review

- One method for determining the distance to a star is to use a measurement called *stellar parallax*, the extremely slight back–and–forth shifting in a nearby star's position due to the orbital motion of Earth. *The farther away a star is, the less its parallax.* A unit used to express stellar distance is the *light-year*, which is the distance light travels in a year—about 9.5 trillion kilometers (5.8 trillion miles).

- The intrinsic properties of stars include *brightness, color, temperature, mass,* and *size.* Three factors control the brightness of a star as seen from Earth: how big it is, how hot it is, and how far away it is. *Magnitude* is the measure of a star's brightness. *Apparent magnitude* is how bright a star appears when viewed from Earth. *Absolute magnitude* is the "true" brightness if a star were at a standard distance of about 32.6 light-years. The difference between the two magnitudes is directly related to a star's distance. Color is a manifestation of a star's temperature. Very hot stars (surface temperatures above 30,000 K) appear blue; red stars are much cooler (surface temperatures generally less than 3000 K). Stars with surface temperatures between 5000 and 6000 K appear yellow, like the sun. The center of mass of orbiting *binary stars* (two stars revolving around a common center of mass under their mutual gravitational attraction) is used to determine the mass of the individual stars in a binary system.

• A *Hertzsprung-Russell diagram* is constructed by plotting the absolute magnitudes and temperatures of stars on a graph. A great deal about the sizes of stars can be learned from H-R diagrams. Stars located in the upper-right position of an H-R diagram are called *giants*, luminous stars of large radius. *Supergiants* are very large. Very small *white dwarf* stars are located in the lower-central portion of an H-R diagram. Ninety percent of all stars, called *main-sequence stars*, are in a band that runs from the upper-left corner to the lower-right corner of an H-R diagram.

• New stars are born out of enormous accumulations of dust and gases, called *nebula*, that are scattered between existing stars. A *bright nebula* glows because the matter is close to a very hot (blue) star. The two main types of bright nebulae are *emission nebulae* (which derive their visible light from the fluorescence of the ultraviolet light from a star in or near the nebula) and *reflection nebulae* (relatively dense dust clouds in interstellar space that are illuminated by reflecting the light of nearby stars). When a nebula is not close enough to a bright star to be illuminated, it is referred to as a *dark nebula.*

• Stars are born when their nuclear furnaces are ignited by the unimaginable pressures and temperatures in collapsing nebulae. New stars not yet hot enough for nuclear fusion are called *protostars.* When collapse causes the core of a protostar to reach a temperature of at least 10 million K, the fusion of hydrogen nuclei into helium nuclei begins in a process called *hydrogen burning.* The opposing forces acting on a star are *gravity* trying to contract it and *gas pressure (thermal nuclear energy)* trying to expand it. When the two forces are balanced, the star becomes a stable *main-sequence star.* When the hydrogen in a star's core is consumed, its outer envelope expands enormously and a *red giant* star, hundreds to thousands of times larger than its main-sequence size, forms. When all the usable nuclear fuel in these giants is exhausted and gravity takes over, the stellar remnant collapses into a small dense body.

• The *final fate of a star is determined by its mass.* Stars with less than one-half the mass of the sun collapse into hot, dense *white dwarf* stars. Medium mass stars (between 0.5 and 3.0 times the mass of the sun) become red giants, collapse, and end up as white dwarf stars, often surrounded by expanding spherical clouds of glowing gas called *planetary nebulae.* Stars more than three times the mass of the sun terminate in a brilliant explosion called a *supernova.* Supernovae events can produce small, extremely dense *neutron stars*, composed entirely of subatomic particles called neutrons; or even smaller and more dense *black holes*, objects that have such immense gravity that light cannot escape their surface.

• The *Milky Way galaxy* is a large, disk-shaped, *spiral galaxy* about 100,000 light-years wide and about 10,000 light-years thick at the center. There are three distinct *spiral arms* of stars, with some showing splintering. The sun is positioned in one of these arms about two-thirds of the way from the galactic center, at a distance of about 30,000 light-years. Surrounding the galactic disk is a nearly spherical halo made of very tenuous gas and numerous *globular clusters* (nearly spherically shaped groups of densely packed stars).

• The various types of galaxies include 1) *irregular galaxies*, which lack symmetry and account for only 10 percent of the known galaxies; 2) *spiral galaxies*, which are typically disk-shaped with a somewhat greater concentration of stars near their centers, often containing arms of stars extending from their central nucleus; and 3) *elliptical galaxies*, the most abundant type, which have an ellipsoidal shape that ranges to nearly spherical, and lack spiral arms.

• By applying the *Doppler effect* (the apparent change in wavelength of radiation caused by the motions of the source and the observer) to the light of galaxies, galactic motion can be determined. Most galaxies have Doppler shifts toward the red end of the spectrum, indicating increasing distance. The amount of Doppler shift is dependent on the velocity at which the object is moving. Because the most distant galaxies have the greatest red shifts, Edwin Hubble concluded in the early 1900s that they were retreating from us with greater recessional velocities than more nearby galaxies. It was soon realized that an *expanding universe* can adequately account for the observed red shifts.

• The belief in the expanding universe led to the widely accepted *Big Bang theory.* According to this theory, the entire universe was at one time confined in a dense, hot, supermassive concentration. About 20 billion years ago, a cataclysmic explosion hurled this material in all directions, creating all matter and space. Eventually the ejected masses of gas cooled and condensed, forming the stellar systems we now observe fleeing from their place of origin.

Chapter Outline

I. Properties of stars
  A. Distance
    1. Measuring can be very difficult
    2. Stellar parallax
      a. Used for measuring distance to a star
      b. Apparent shift in a star's position due to orbital motion of Earth
      c. Measured as an angle
      d. Near stars have largest parallax—largest less than 1 second of arc
    3. Distances to stars very large
    4. Units of measurement
      a. Kilometers or astronomical units are too cumbersome to use
      b. Light-year
        1. Distance light travels in one year
        2. 9.5 trillion km (5.8 trillion miles)
    5. Other methods for measuring also used
  B. Stellar brightness
    1. Controlled by three factors
      a. Size
      b. Temperature
      c. Distance
    2. Magnitude
      a. Measure of a star's brightness
      b. Two types
        1. Apparent magnitude
          a. Brightness viewed from Earth
          b. Decreases with distance
          c. Numbers used to designate
            1. Dim stars, large numbers
              a. First magnitude stars appear bright
              b. Sixth magnitude stars are faintest visible to the eye
            2. Negative numbers also used
        2. Absolute magnitude
          a. "True" or intrinsic brightness
          b. Brightness at a standard distance of 32.6 light-years
          c. Most stars between -5 and +15
  C. Color and temperature
    1. Hot star
      a. Temperature above 30,000 K
      b. Emits short-wavelength light
      c. Appears blue
    2. Cool star
      a. Temperature less than 3000 K
      b. Emits longer-wavelength light
      c. Appears red
    3. Between 5000 and 6000 K
      a. Stars appear yellow
      b. e.g., Sun
  D. Binary stars and stellar mass
    1. Binary stars
      a. Two stars orbiting one another
        1. Held together by mutual gravitation
        2. Both orbit around a center of mass
      b. Visual binaries can be resolved telescopically
      c. More than 50% of stars in universe
      d. Used to determine mass

2. Stellar mass
   a. Determined using binary stars
   b. Mass of most stars is between one-tenth and fifty times the mass of the sun

II. Hertzsprung-Russell diagram
  A. Shows relation between stellar brightness (absolute magnitude) and temperature
  B. Diagram made by plotting (graphing) each star's
    1. Luminosity (brightness) and
    2. Temperature
  C. Parts of H-R diagram
    1. Main-sequence stars
      a. 90% of all stars
      b. Band through center of H-R diagram
      c. Sun in main-sequence
    2. Giants (or red giants)
      a. Very luminous
      b. Large
      c. Upper-right on H-R diagram
      d. Very large giants are called supergiants
      e. Only a few percent of all stars
    3. White dwarfs
      a. Fainter than main-sequence stars
      b. Small—some approximate the size of Earth
      c. Lower-central on H-R diagram
      d. Not all are white
      e. Perhaps 10% of all stars
  D. Used to study stellar evolution

III. Interstellar matter
  A. Between stars is "the vacuum of space"
  B. Nebula
    1. Cloud of dust and gases
    2. Two major types
      a. Bright nebula
        1. Glows if close to very hot star
        2. Two types
          a. Emission nebula
            1. Largely hydrogen
            2. Absorb ultraviolet radiation
            3. Emit visible light
          b. Reflection nebula
            1. Reflect light of nearby stars
            2. Interstellar dust
      b. Dark nebula
        1. Not close to bright star
        2. Appear dark
    3. Material that forms stars and planets

IV. Stellar evolution
  A. Stars exist because of gravity
  B. Two opposing forces in a star
    1. Gravity contracts
    2. Thermal nuclear energy expands
  C. Stages
    1. Birth
      a. In dark, cool, interstellar clouds
      b. Gravity contracts cloud
      c. Temperature rises
      d. Radiates long-wavelength red light
      e. Becomes protostar
    2. Protostar
      a. Gravitational contraction continues
      b. Core reaches 10 million K
      c. Hydrogen nuclei fuse
        1. Become helium nuclei
        2. Process called hydrogen burning
      d. Energy released
      e. Outward pressure increases
      f. Outward pressure balanced by gravity
      g. Star becomes a stable main-sequence star
    3. Main-sequence stage
      a. Stars age at different rates
        1. Massive stars use fuel faster and live a few million years
        2. Small stars use fuel more slowly and live perhaps hundreds of billions of years
      b. 90% of star's life is in main-sequence
    4. Red giant stage
      a. Hydrogen burning migrates outward
      b. Star's outer envelope expands
        1. Surface cools
        2. Becomes red
      c. Core collapsing—helium converted to carbon
      d. Eventually all nuclear fuel is used
      e. Gravity squeezes star
    5. Burnout and death
      a. Final stage depends on mass
      b. Possibilities
        1. Low-mass star
          a. 0.5 solar mass
          b. Red giant collapses
          c. Becomes white dwarf

2. Medium-mass
   a. Between 0.5 and 3 solar masses
   b. Red giant collapses
   c. Planetary nebula forms
      1. Cloud of gas
      2. Outer layer of star
   d. Becomes white dwarf
3. Massive star
   a. Over 3 solar masses
   b. Short life span
   c. Terminates in a brilliant explosion called a supernova
   d. Interior condenses
   e. May produce hot, dense object
      1. Neutron star, or
      2. Black hole

D. H-R diagrams used to study stellar evolution

V. Stellar remnants
   A. White dwarf
      1. Small—some no larger than Earth
      2. Dense
         a. Can be more massive than sun
         b. Spoonful weighs several tons
         c. Atoms take up less space
            1. Electrons displaced inward
            2. Called degenerate matter
            3. Hot surface
            4. Cools to become black dwarf
   B. Neutron star
      1. Forms from more massive star
         a. Has more gravity
         b. Squeezes itself smaller
      2. Remnant of supernova
      3. Gravitational force collapses atoms and electrons combine with protons to produce neutrons
      4. Pea size sample
         a. Weighs 100 million tons
         b. Same density as atomic nucleus
      5. Strong magnetic field
      6. First one discovered in early 1970s
         a. Pulsar—pulsating radio source
         b. Found in Crab nebula—remnant of A.D. 1054 supernova
   C. Black hole
      1. More dense than neutron star
      2. Intense surface gravity—no light can escape
      3. As matter is pulled into it
         a. Becomes very hot
         b. Emits x-rays
      4. Likely candidate is Cygnus X-1, a strong x-ray source

VI. Galaxies
   A. Milky Way galaxy
      1. Structure
         a. Determined using radio telescopes
         b. Large spiral galaxy
            1. About 100,000 light-years wide
            2. Thickness at the galactic nucleus is about 10,000 light-years
         c. Three spiral arms
         d. Sun is 30,000 light-years from center
      2. Rotation
         a. Around galactic nucleus
         b. Outermost stars move the slowest
         c. Sun rotates around galactic nucleus
         d. Takes about 200 million years
      3. Halo surrounds galactic disk
         a. Spherical
         b. Very tenuous gas
         c. Numerous globular clusters
   B. Other galaxies
      1. Existence first proposed in mid-1700s by Immanuel Kant
      2. Three basic types
         a. Irregular galaxy
            1. Lacks symmetry
            2. About 10 percent of all galaxies
            3. Mostly young stars
            4. Examples are Magellanic Clouds
         b. Spiral galaxy
            1. Arms extending from nucleus
            2. About 30 percent of all galaxies
            3. Large diameter—20,000 to 125,000 light-years
            4. Contains both young and old stars
            5. e.g., Milky Way
         c. Elliptical galaxy
            1. Ellipsoidal shape
            2. About 60 percent of all galaxies
            3. Size
               a. Most are smaller than spiral galaxies
               b. Also largest known galaxies
   C. Galactic cluster
      1. Group of galaxies
      2. Some contain thousands of galaxies
      3. Local Group
         a. Our own group of galaxies

b. At least 28 galaxies
4. Supercluster
a. Huge swarm of galaxies
b. May be largest entity in universe

VII. Red shifts
A. Doppler effect
1. Change in wavelength due to motion
2. With light
a. Movement away stretches wavelength
1. Longer wavelength
2. Light appears redder
b. Movement toward squeezes wavelength
1. Shorter wavelength
2. Light shifted toward the blue
3. Amount of shift indicates rate
a. Large Doppler shift indicates a high velocity
b. Small Doppler shift indicates a lower velocity
B. Expanding universe
1. Most galaxies exhibit red shift
b. Far galaxies exhibit the greatest shift greater velocity
2. Discovered
a. 1929
b. Edwin Hubble
3. Hubble's Law—recessional speed of galaxies is proportional to their distance
4. Accounts for red shifts

VIII. Big Bang theory
A. Accounts for galaxies moving away from us
B. Universe once confined to a "ball"
1. Supermassive
2. Dense
3. Hot
C. Big bang marks inception of the universe
1. About 20 billion years ago
2. All matter and space created
D. Matter moving outward
E. Outward expansion could stop
1. If concentration of matter is great enough
2. Gravitational contraction

Vocabulary Review

Choosing from the list of key terms, furnish the most appropriate response for the following statements.

1. The measurement of a star's brightness is called its ____________.

2. A(n) ____________ is a type of galaxy that lacks symmetry.

3. A(n) ____________ is a small, dense star that was once a low-mass or medium-mass star whose internal heat energy was able to keep these gaseous bodies from collapsing under their own weight.

4. A graph showing the relation between the true brightness and temperature of stars is called a(n) ____________.

5. The fusion of groups of four hydrogen nuclei into a single helium nuclei is called ____________.

6. ____________ is a type of very dense matter that forms when electrons are displaced inward from their regular orbits around an atom's nucleus.

Key Terms

Page numbers shown in ( ) refer to the textbook page where the term first appears.

absolute magnitude (p. 414)
apparent magnitude (p. 413)
Big Bang (p. 430)
black hole (p. 425)
bright nebula (p. 417)
dark nebula (p. 418)
degenerate matter (p. 423)
Doppler effect (p. 429)
elliptical galaxy (p. 427)
emission nebula (p. 417)
galactic cluster (p. 428)
giant (p. 416)
Hertzsprung-Russell (H-R) diagram (p. 435)
Hubble's law (p. 430)
hydrogen burning (p. 419)
interstellar dust (p. 418)
irregular galaxy (p. 428)
light-year (p. 412)
Local Group (p. 428)
magnitude (p. 413)
main-sequence stars (p. 415)
nebula (p. 417)
neutron star (p. 424)
planetary nebula (p. 422)
protostar (p. 419)
pulsar (p. 424)
red giant (p. 416)
reflection nebula (p. 418)
spiral galaxy (p. 427)
stellar parallax (p. 412)
supergiant (p. 436)
supernova (p. 422)
white dwarf (p. 417)

7. ____________ is the extremely slight back-and-forth shifting in the position of a nearby star due to the orbital motion of Earth.

8. An interstellar cloud consisting of dust and gases is referred to as a(n) ____________.

9. A(n) ____________ is an exceptionally large star, such as Betelgeuse, with a radius about 800 times that of the sun and high luminosity .

10. A(n) ____________ is a gaseous interstellar mass that absorbs ultraviolet light from an embedded or nearby hot star and reradiates, or emits, this energy as visible light.

11. A collapsing cloud of gases and dust not yet hot enough to engage in nuclear fusion but destined to become a star is referred to as a(n) ____________.

12. A star consisting of matter formed from the combination of electrons with protons that is smaller and more massive than a white dwarf is called a(n) ____________.

13. A(n) ____________ is a very large, cool, reddish colored star of high luminosity.

14. Stars exceeding 3 solar masses have relatively short life spans and terminate in a brilliant explosion called a(n) ____________.

15. When a dense cloud of interstellar material is not close enough to a bright star to be illuminated, it is referred to as a(n) ____________.

16. Discovered in the early 1970s, a(n) ____________ is a neutron star that radiates short bursts of radio energy.

17. ____________ states that galaxies are receding from us at a speed that is proportional to their distance.

18. A star's brightness as it appears when viewed from Earth is referred to as its ____________.

19. A type of galaxy that has an ellipsoidal shape that ranges to nearly spherical and lacks spiral arms is called a(n) ____________.

20. A(n) ____________ is a unit used to express stellar distance equal to about 9.5 trillion kilometers (5.8 trillion miles).

21. A(n) ____________ is an interstellar cloud that glows because of its proximity to a very hot (blue) star.

22. An object even smaller and denser than a neutron star, with a surface gravity so immense that light cannot escape, is called a(n) ____________.

23. An interstellar cloud of gases and dust that merely reflects the light of nearby stars is called a(n) ____________.

24. The Milky Way is a type of galaxy called a(n) ____________.

25. The brightness of a star, if it were viewed at a distance of 32.6 light-years, is called its ____________.

26. A reflection nebula is thought to be composed of a rather dense cloud of large particles called ____________.

27. A(n) ____________ is a very luminous star with a large radius.

28. On an H-R diagram, the "ordinary" stars that are located along a band that extends from the upper-left corner to the lower–right corner are called ____________.

29. An expanding spherical cloud of gas, called a(n) ____________, forms when a medium-mass star collapses from a red giant to a white dwarf star.

30. The theory which proposes that the universe originated as a single mass that subsequently exploded is referred to as the ____________ theory.

31. Our own group of galaxies, called the ____________, contains at least 28 galaxies.

32. The apparent change in wavelength of radiation caused by the relative motions of the source and the observer is called the ____________.

33. A system of galaxies containing from several to thousands of member galaxies is called a(n) ____________.

Comprehensive Review

1. Briefly describe the relation between a star's distance and its parallax.

2. What causes the difference between a star's apparent magnitude and its absolute magnitude?

3. List the three factors that control the apparent brightness of a star as seen from Earth.

   1)

   2)

   3)

4. What is the approximate surface temperature of a star with the following color?

   a) Red:

   b) Yellow:

   c) Blue:

5. Describe how binary stars are used to determine stellar mass.

6. List the two properties of stars that are used to construct an H-R diagram.

   1)

   2)

7. Using Figure 16.1, an idealized Hertzsprung-Russell diagram, select the letter that indicates the location of each of the following types of stars.

   a) White dwarfs: ____ c) Red stars (lowmass): ____

   b) Blue stars: ____ d) Giants: ____

8. On Figure 16.1, which letter is most representative of our sun? ____

9. On Figure 16.1, which stars are on the main sequence?

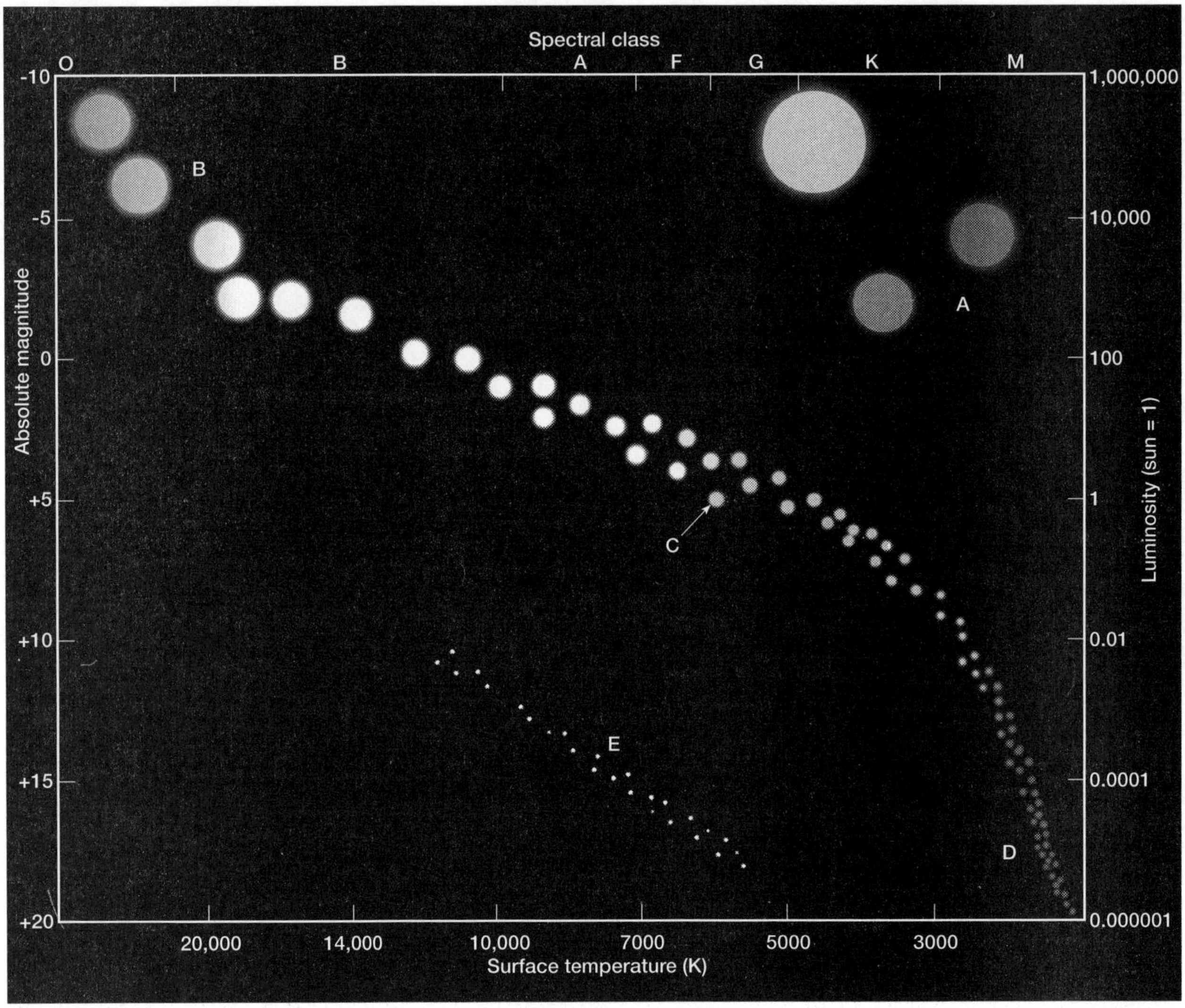

Figure 16.1

10. List and briefly describe the two main types of bright nebulae.
    1)

    2)

11. What circumstance produces a dark nebula?

12. Describe the process that astronomers refer to as hydrogen burning.

13. What will be the most likely final stage in the burnout and death of stars in each of the following mass categories?

    a) Low-mass stars:

    b) Medium-mass (sunlike) stars:

    c) Massive stars:

14. On Figure 16.2, beginning with dust and gases, draw a sequence of arrows that show in correct order the stages that a star about as massive as the sun will follow in its evolution.

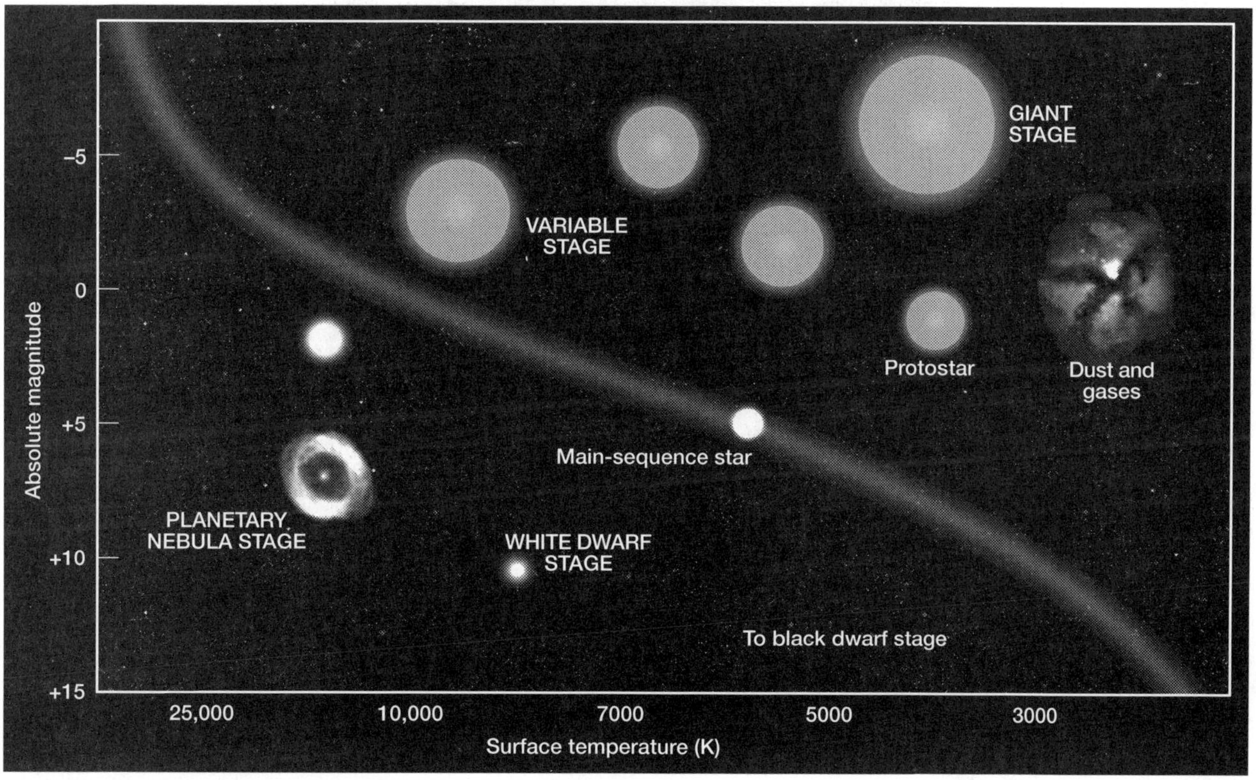

Figure 16.2

15. Why are radio telescopes rather than optical telescopes used to determine the gross structure of the Milky Way galaxy?

16. Describe the size and structure of the Milky Way.

17. List and describe the three basic types of galaxies.
    1)

    2)

    3)

18. The large galaxy shown in Figure 16.3 is of which type?

Figure 16.3

19. How is the Doppler effect used to determine whether Earth and another celestial body are approaching or leaving one another and the rate at which the relative movement is occurring?

20. What is Hubble's law?

21. Briefly describe the Big Bang theory.

Practice Test

Multiple choice. Choose the best answer for the following multiple choice questions.

1. The two properties of a star that are plotted on an H-R diagram are ___________.
   a) size and distance
   b) brightness and temperature
   c) speed and distance
   d) size and temperature
   e) brightness and distance

2. Stars with surface temperatures between 5000 and 6000 K appear ___________.
   a) red
   b) green
   c) violet
   d) yellow
   e) blue

3. A star with which one of the following magnitudes would appear the brightest?
   a) 15
   b) 10
   c) 5
   d) 1
   e) -5

4. Which force is most responsible for the formation of a star?
   a) magnetic
   b) gravity
   c) nuclear
   d) interstellar
   e) light pressure

5. The difference in the brightness of two stars having the same surface temperature is attributable to their relative ___________.
   a) densities
   b) colors
   c) compositions
   d) ages
   e) sizes

6. In the cores of incredibly hot red giant stars, nuclear reactions convert helium to ___________.
   a) carbon
   b) hydrogen
   c) oxygen
   d) argon
   e) nitrogen

7. Stellar parallax is used to determine a star's ___________.
   a) motion
   b) brightness
   c) distance
   d) mass
   e) temperature

8. During the process referred to as hydrogen burning, hydrogen nuclei in the core of a star are fused together and become ___________ nuclei.
   a) oxygen
   b) nitrogen
   c) helium
   d) carbon
   e) argon

9. Binary stars can be used to determine stellar ___________.
   a) mass
   b) temperature
   c) brightness
   d) distance
   e) composition

10. One light-year is about ___________.
    a) 6 trillion kilometers
    b) 9.5 trillion miles
    c) 8 trillion kilometers
    d) 5.8 trillion miles
    e) 11 trillion kilometers

11. Which one of the following stellar remnants has the greatest density?
    a) neutron star b) black hole c) white dwarf d) black dwarf

12. An interstellar accumulation of dust and gases is referred to as a ____________.
    a) red giant c) white dwarf e) neutron star
    b) nebula d) galaxy

13. Which one of the following is NOT a type of galaxy?
    a) nebular c) spiral e) barred spiral
    b) irregular d) elliptical

14. The most abundant gas in dark, cool, interstellar clouds in the neighborhood of the Milky Way is ____________.
    a) hydrogen c) helium e) nitrogen
    b) argon d) oxygen

15. A very massive star terminates in a brilliant explosion called a ____________.
    a) red giant c) protostar e) planetary nebula
    b) supernova d) neutron star

16. According to the Big Bang theory, a cataclysmic explosion created the matter in the universe about ____________ years ago.
    a) 600 million c) 4.6 billion e) 20 billion
    b) 2 billion d) 12 billion

17. Which one of the following is NOT a type of nebula?
    a) reflection b) emission c) spiral d) dark

18. The apparent change in the position of a celestial object due to the orbital motion of Earth is called ____________.
    a) parallax c) magnitude e) orbital shifting
    b) recessional velocity d) brightness

19. Matter that is pulled into a black hole should become very hot and emit ____________ before being engulfed.
    a) x-rays c) atoms e) infrared radiation
    b) hydrogen nuclei d) degenerate matter

20. Which one of the following stars would appear brightest in the night sky?
    a) Sirius c) Alpha Centauri e) Betelgeuse
    b) Deneb d) Arcturus

21. The Milky Way is a ____________.
    a) constellation c) spiral galaxy e) galactic cluster
    b) red giant d) dark nebula

22. On an H-R diagram, main-sequence blue stars are massive and ____________.
    a) cool c) hot e) supergiants
    b) white dwarfs d) novae

23. The measurement of a star's brightness is called ____________.
    a) temperature c) parallax e) pulsation
    b) magnitude d) period

24. A very young, large, red object that is soon to become a star, but is not yet hot enough for nuclear fusion, is termed a(n) ____________.
    a) black hole c) red giant e) white dwarf
    b) supernova d) protostar

25. The apparent change in wavelength of radiation caused by the relative motions of the source and the observer is referred to as ____________.
    a) Hubble's law d) relativity
    b) the acceleration factor e) the Doppler effect
    c) Newton's law of radiation

26. The density of a white dwarf is such that a spoonful of its matter would weigh ____________.
    a) a kilogram c) 40 grams e) several tons
    b) 500 pounds d) a ton

27. The sun is positioned about ____________ of the way from the center of the galaxy.
    a) one-fourth c) one-half e) three-fourths
    b) one-third d) two-thirds

28. The brightness of a star, if it were at a distance of 32.6 light-years, is its ____________.
    a) apparent magnitude c) apparent brightness e) mass-distance ratio
    b) parallax d) absolute magnitude

29. Which type of galaxy is composed mostly of young stars?
    a) barred spiral c) irregular e) nebular
    b) elliptical d) spiral

30. The red shifts exhibited by galaxies can be explained by ____________.
    a) calculation errors d) expansion of the Milky Way
    b) an expanding universe e) a collapsing universe
    c) Kepler's first law

31. On an H-R diagram, about 90 percent of the stars are ____________.
    a) giants c) white dwarfs e) novae
    b) main-sequence stars d) super giants

32. The difference between a star's apparent and absolute magnitudes is related to ____________.
    a) surface temperature c) core temperature e) distance
    b) color d) time

True/false. For the following true/false questions, if a statement is not completely true, mark it false. For each false statement, change the ***italicized*** *word to correct the statement.*

1. ___ A hot, massive blue star will age more *slowly* than a small (red) main sequence star.
2. ___ The parallax of a far star is *less* than that of a near star.
3. ___ A stable main-sequence star is balanced between the force of gravity and internal *gas* pressure
4. ___ Galaxies are receding from us at a speed that is proportional to their *size.*
5. ___ *Absolute* magnitude is the brightness of a star when viewed from Earth.
6. ___ Stars with equal surface *temperatures* radiate the same amount of energy per unit area.
7. ___ On an H-R diagram, the sun is a *main-sequence* star.
8. ___ The larger the magnitude number, the *dimmer* the star.
9. ___ *Low-mass* stars evolve to become white dwarfs.
10. ___ On an H-R diagram, the hottest main-sequence stars are intrinsically the *brightest.*
11. ___ A star's color is primarily a manifestation of its *temperature.*
12. ___ Giant stars are *cooler* than white dwarfs.
13. ___ *Dark* nebulae glow because they are close to very hot stars.
14. ___ The light-year is a unit used to measure the *age* of a star.
15. ___ All stars eventually exhaust their nuclear fuel and collapse in response to *gravity.*
16. ___ A star with a surface temperature less than 3000 K will generally appear *red.*
17. ___ A rapidly rotating neutron star that radiates short pulses of radio energy is called a *pulsar.*
18. ___ The Milky Way is about 100,000 *kilometers* wide.
19. ___ *Spiral* galaxies are the most abundant type of galaxies.
20. ___ A first-magnitude star is about 100 times *dimmer* than a sixth-magnitude star.
21. ___ Large Doppler shifts indicate *low* velocities.
22. ___ Stars that orbit one another are called *neutron* stars.
23. ___ If there is enough *matter* in the universe, the outward expansion of the galaxies could eventually stop.

Written questions

1. Describe the movement of an average star like the sun through the H-R diagram during its lifetime.

2. What is the final state for 1) a low-mass (red) main sequence star, 2) a medium-mass (sun-like) star, and 3) a very massive star?

3. Describe the Big Bang theory.

INTRODUCTION

Vocabulary Review

1. core, mantle, crust	5. resource	9. renewable resource	13. environment
2. hypothesise	6. theory	10. biosphere	14. Astronomy
3. geology	7. hydrosphere	11. nonrenewable resource	15. atmosphere
4. meteorology	8. oceanography	12. Earth science	16. law

Comprehensive Review

1. The sciences traditionally included in earth science are geology, oceanography, meteorology, and astronomy.

2. Renewable resources, such as forest products and wind energy, can be replenished over relatively short time spans. By contrast, nonrenewable resources, such as oil and copper, form so slowly that significant deposits often take millions of years to accumulate. Therefore, practically speaking, the Earth contains fixed quantities of these materials.

3. 1) hydrosphere; a dynamic mass of water that is continually on the move from the oceans to the atmosphere, precipitating back to the land, and returning back to the ocean again; 2) atmosphere; Earth's blanket of air; 3) solid Earth; which includes the core, mantle, and crust; 4) biosphere; which includes all life on Earth

4. Earth science is the science which collectively seeks to understand Earth and its neighbors in space. It includes geology, oceanography, meteorology, and astronomy.

5. A theory is a well-tested and widely accepted view that best explains certain observable facts; while a scientific hypothesis is an untested explanation.

6. 1) collection of facts (data) through observation and measurement; 2) development of a working hypothesis to explain the facts; 3) construction of experiments to validate the hypothesis; 4) acceptance, modification, or rejection of the hypothesis on the basis of extensive testing

7. A (inner core); B (outer core); C (mantle); D (crust)

8 a) Physical geology studies the materials that compose Earth and seeks to understand the many processes that operate beneath and upon its surface. b) Historical geology seeks to understand the origin of Earth and the development of the planet through its 4.6-billion-year history.

Practice Test

Multiple choice

1. d	3. c	5. d	7. b	9. d	11. d	13. c	15. a	17. c
2. d	4. a	6. b	8. a	10. a	12. d	14. b	16. d	

True/false

1. T
2. F (hydrosphere)
3. F (nonrenewable)
4. F (historical)
5. F (predictable)
6. F (environment)
7. T
8. F (crust)
9. T
10. T

Written questions

1. The four steps that scientists often use to conduct experiments and gain scientific knowledge are 1) collection of facts (data) through observation and measurement; 2) development of a working hypothesis to explain the facts; 3) construction of experiments to validate the hypothesis; and 4) acceptance, modification, or rejection of the hypothesis on the basis of extensive testing.
2. Earth's four "spheres" include the 1) hydrosphere, a dynamic mass of water that is continually on the move from the oceans to the atmosphere, precipitating back to the land, and returning back to the ocean again; 2) atmosphere, Earth's blanket of air; 3) solid Earth, which includes the core, mantle, and crust; and 4) biosphere, which includes all life on Earth.
3. Earth science is the science which collectively seeks to understand Earth and its neighbors in space. It includes geology, oceanography, meteorology, and astronomy.

CHAPTER ONE

Vocabulary Review

1. neutron
2. atom
3. ore
4. mineral
5. ion
6. cleavage
7. proton
8. compound
9. silicates
10. reserve
11. electron
12. rock
13. nucleus
14. atomic number
15. element
16. mass number
17. energy level
18. isotope
19. crystal form
20. radioactivity
21. luster
22. silicon-oxygen tetrahedron
23. mineral resource
24. streak
25. hardness
26. fracture
27. specific gravity
28. Mohs hardness scale
29. color

Comprehensive Review

1. 1) It must be naturally occurring. 2) It must be inorganic. 3) It must be solid. 4) It must possess a definite chemical structure.
2. 1) protons; particles with positive electrical charges that are found in an atom's nucleus (Letter: C); 2) neutrons; particles with neutral electrical charges that are found in an atom's nucleus (Letter: B); 3) electrons; negative electrical charges surrounding the atom's nucleus in energy levels (Letter: A)
3. Minerals are the same throughout; while most rocks are mixtures, or aggregates, of minerals where each mineral retains its distinctive properties.
4. a) 6; b) 14
5. a) 17; b) 17; c) 18
6. The sodium atom loses one electron and becomes a positive ion. The chlorine atom gains one electron and becomes a negative ion. These oppositely charged ions attract one another and the bond produces the compound sodium chloride.
7. Yes, the atom is an ion because it contains two more electrons (10) than protons (8). a) 13; b) 8
8. No, the atom contains an equal number of protons (17) and electrons (17) and is neutral.
9. A mineral is a naturally occurring inorganic solid that possesses a definite chemical structure.
10. An isotope will have a different mass number due to a different number of neutrons.
11. a) the appearance of light reflected from the surface of a mineral; b) the external expression of a mineral's orderly arrangement of atoms; c) the color of a mineral in its powdered form; d) a measure of the resistance of a mineral to abrasion or scratching; e) the tendency of a mineral to break along planes of weak bonding; f) the uneven, or irregular, breaking of a mineral; minerals that do not exhibit cleavage when broken are said to fracture; g) the weight of a mineral compared to the weight of an equal volume of water

12. oxygen (symbol: O, 46.6%) and silicon (symbol: Si, 27.7%)
13. letter A (oxygen); letter B (silicon)
14. a) feldspars, quartz; b) micas (biotite, muscovite); c) amphibole group; d) pyroxene group; e) olivine
15. a) feldspar, quartz, and muscovite; b) calcite
16. A rock is an aggregate of minerals.
17. Most silicates crystallize from molten rock as it cools.
18. a) cinnabar; b) galena; c) sphalerite; d) hematite (see Table 1.3)
19. As an isotope decays through a process called radioactivity, it actively radiates energy and particles. The decay often produces a different isotope of the same element, but with a different mass number.
20. No. Before aluminum can be extracted profitably it must be concentrated to four times its average crustal percentage of 8.13 percent. The sample, with only 10 percent aluminum, lacks the necessary concentration.

Practice Test

Multiple choice

1. c	6. a	11. d	16. c	21. b
2. b	7. c	12. b	17. a	22. e
3. e	8. c	13. a	18. b	23. a
4. c	9. c	14. e	19. e	24. a
5. a	10. b	15. b	20. d	25. c

True/false

1. F (electrons)	10. T	19. F (compound)
2. T	11. F (one-fourth)	20. T
3. F (four-thousand)	12. F (carbonate)	21. F (electrons)
4. T	13. F (gypsum)	22. T
5. F (oxygen)	14. T	23. F (atom)
6. T	15. F (crystal form)	24. T
7. T	16. T	25. F (harder)
8. T	17. F (nonmetallic)	
9. F (between)	18. T	

Written questions

1. Protons and neutrons are found in the atom's nucleus and represent practically all of an atom's mass. Protons are positively charged, while neutrons are neutral. Electrons orbit the nucleus in regions called energy-levels and carry a negative charge.
2. Most mineral crystals don't visibly demonstrate their crystal form because, most of the time, crystal growth is severely constrained because of space restrictions where, and when, the mineral forms.
3. The basic building block of all silicate minerals is the silicon-oxygen tetrahedron. It consists of four oxygen atoms surrounding a much smaller silicon atom.
4. For any material to be considered a mineral it must be 1) naturally occurring, 2) inorganic, 3) solid, and 4) possess a definite chemical structure.
5. A mineral resource is any useful mineral that can be recovered for use. Resources include already identified deposits from which minerals can be extracted profitably, called reserves.

CHAPTER TWO

Vocabulary Review

1. rock cycle
2. weathering
3. lithification
4. metamorphic rock
5. magma
6. crystallization
7. regional metamorphism
8. sedimentary rock
9. lava
10. texture
11. foliated texture
12. extrusive (volcanic)
13. sediment
14. igneous rock
15. foliated texture
16. fine-grained texture
17. frost wedging
18. chemical sedimentary rock
19. porphyritic texture
20. strata (beds)
21. evaporite deposit
22. contact metamorphism
23. glassy texture
24. detrital sedimentary rock
25. coarse-grained texture
26. exfoliation dome

Comprehensive Review

1. A (Crystallization); B (Igneous); C (Weathering, transportation, and deposition); D (Sediment); E (Lithification); F (Sedimentary); G (Metamorphism); H (Metamorphic); I (Melting)

2. James Hutton; late 1700s

3. any other

4. running water, wind, waves, and glacial ice

5. a) Igneous rocks form as magma cools and solidifies either beneath or at Earth's surface. b) Sedimentary rocks are the lithified products of weathering. c) Metamorphic rocks can form from igneous, sedimentary, or other metamorphic rocks that are subjected to heat, pressure, and chemically active fluids.

6. The rate of cooling most influences the size of mineral crystals in igneous rock. Slow cooling results in the formation of large crystals; while rapid cooling often results in a mass of very small intergrown crystals.

7. Magma is molten rock beneath the surface. Lava, molten rock on the surface, is similar to magma except that most of the gaseous component has escaped.

8. 1) texture; the overall appearance of an igneous rock based on the size and arrangement of its interlocking crystals; 2) mineral composition; the mineral makeup of an igneous rock which depends on the composition of the magma from which it originated

9. a) rapid cooling at the surface or as small masses within the upper crust (letter A.); b) very rapid cooling, perhaps when molten rock is ejected into the atmosphere (letter D.); c) slow cooling of large masses of magma far below the surface (letter B.); d) two different rates of cooling, resulting in two different crystal sizes (letter C.)

10. N. L. Bowen discovered that, as magma cools in the laboratory, certain mineral crystallize first at very high temperatures. At successively lower temperatures, other minerals crystallize.
11. olivine; pyroxene
12. Yes, providing that they have different textures.
13. Weathering is the natural response of Earth materials to a changing environment.
14. 1) Mechanical weathering involves breaking a rock into smaller and smaller pieces. 2) Chemical weathering alters the internal structure of minerals by removing and/or adding elements.
15. 1) Frost wedging breaks rock into smaller fragments by the alternate freezing and thawing of water. 2) Expansion from unloading occurs when large masses of igneous rock are exposed by erosion, expand, and break loose, like the layers of an onion. 3) Weathering by biological activity is accomplished by the activities of organisms, including plants, burrowing animals, and humans.
16. clay minerals and dissolved silica (both from feldspar), quartz grains
17. compaction and cementation
18. a) Detrital sedimentary rocks are made of a wide variety of mineral and rock fragments, clay minerals and quartz dominate (Example: sandstone). b) Chemical sedimentary rocks are derived from material that is carried in solution to lakes and seas and, when conditions are right, precipitates to form chemical sediments (Example: limestone).
19. particle size
20. Angular fragments indicate that the particles were not transported very far from their source prior to deposition.
21. mineral composition
22. Limestone is the most abundant chemical sedimentary rock. Ninety percent of limestone is biochemical sediment. The rest precipitates directly from seawater.
23. swamp environment, peat, lignite, bituminous, anthracite
24. a) detrital; b) breccia; c) C
25. Fossils are traces or remains of prehistoric life. They are useful for interpreting past environments, as time indicators, and for matching up rocks from different places that are the same age.
26. a) Regional metamorphism occurs during mountain building when great quantities of rock are subjected to intense stress and deformation. b) Contact metamorphism occurs when rock is in contact with, or near, a mass of magma.
27. The three agents of metamorphism include heat, pressure, and chemically active fluids.
28. a) A foliated texture results from the alignment of mineral crystals with a preferred orientation, giving a metamorphic rock a layered or banded appearance (Example: slate). b) The minerals in a nonfoliated metamorphic rock are generally equidimensional and not aligned (Example: marble).
29. foliated (gneiss)

Practice Test

Multiple choice

1. b	6. b	11. b	16. d	21. c
2. b	7. a	12. b	17. b	22. b
3. e	8. a	13. a	18. a	23. d
4. d	9. d	14. d	19. d	24. a
5. a	10. c	15. d	20. b	25. a

True/false

1. T
2. F (basalt)
3. T
4. T
5. T
6. T
7. F (sedimentary)
8. F (metamorphic)
9. F (chemical)
10. F (less)
11. T
12. T
13. F (crystallization)
14. T
15. F (metamorphism)
16. F (lava)
17. T
18. T
19. F (sedimentary)
20. T
21. T
22. T
23. F (chemical)
24. T
25. F (large)

Written questions

1. Magma, the material of igneous rocks, is produced when any rock is melted. Sedimentary rocks are formed from the weathered products of any pre-existing rock—igneous sedimentary or metamorphic. Metamorphic rocks are created when any rock type undergoes metamorphism.

2. Bowen's reaction series illustrates the sequence in which minerals crystallize from a magma. It can be used to help explain the mineral makeup of an igneous rock in that minerals that crystallize at about the same time (temperature) are most often found together in the same igneous rock.

3. Quartz and clay minerals are the chief constituents of detrital sedimentary rocks. Clay minerals are the most abundant product of chemical weathering. Quartz is an abundant sediment because it is very resistant to chemical weathering.

4. Mechanical weathering increases the amount of surface area available for chemical attack.

5. Metamorphic processes cause many changes in existing rocks, including increased density, growth of larger crystals, foliation, and the transformation of low-temperature minerals into high-temperature minerals. Furthermore, the introduction of ions generates new minerals, some of which are economically important.

CHAPTER THREE

Vocabulary Review

1. Mass wasting
2. drainage basin
3. zone of saturation
4. porosity
5. Gradient
6. alluvium
7. capacity
8. groundwater
9. yazoo tributary
10. transpiration
11. meander
12. water table
13. discharge
14. artesian well
15. natural levee
16. permeability
17. water cycle
18. floodplain
19. competence
20. divide
21. aquitard
22. Base level
23. spring
24. aquifer
25. suspended load
26. karst topography

Comprehensive Review

1. Mass wasting is the downslope movement of rock and soil under the direct influence of gravity. The factors that play an important role in the process include water, which destroys the cohesion between particles, and the oversteepening of slopes by the activities of people or the undercutting of a valley wall by a stream.
2. A (slump); B (rockslide); C (mudflow); D (earthflow)
3. mass wasting and running water
4. a) C; b) B; c) A; d) D
5. a) the distance that water travels in a unit of time; b) the slope of a stream channel expressed as the vertical drop of a stream over a specified distance; c) the volume of water flowing past a certain point in a given unit of time
6. 4 meters/kilometer (40 meters divided by 10 kilometers)
7. a) decreases; b) increases; c) decreases; d) increases; e) increase
8. a) Ultimate base level is sea level, the lowest level to which stream erosion could lower the land. b) Temporary base levels include lakes and main streams which act as base level for the streams and tributaries entering them.
9. 1) as dissolved load; brought to a stream by groundwater; 2) as suspended load; sand, silt, and clay carried in the water; 3) as bed load; coarser particles moving along the bottom of a stream
10. During a flood, the increase in discharge of a stream results in a greater capacity and the increase in velocity results in greater competence.
11. a) C; b) B; c) A; d) D

12. 1) A narrow valley is typically V-shaped because the primary work of the stream has been downcutting toward base level. The features of narrow valleys include rapids and waterfalls. 2) In a wide valley the stream's energy is directed from side-to-side and downward erosion becomes less dominant. The features of wide valleys include meanders, floodplains, bars, cutoffs, and oxbow lakes.
13. radial pattern; where streams diverge from a central area like the spokes of a wheel
14. a) B; b) C; c) A
15. a) downcutting, rapids, waterfalls, V-shaped, no floodplain, straight course; b) lateral erosion, meanders, floodplain beginning, low gradient; c) large floodplain, widespread meanders, natural levees, back swamps, yazoo tributaries
16. a) D; b) B; c) A; d) C; e) E
17. The shape of the water table is usually a subdued replica of the surface, reaching its highest elevations beneath hills and decreasing in height toward valleys.
18. The source of heat for most hot springs and geysers is in cooling igneous rock beneath the surface.
19. A rock layer or sediment that transmits water freely because of its high porosity and high permeability is called an aquifer.
20. 1) land subsidence caused by groundwater withdrawal; 2) groundwater contamination
21. 1) slowly, over many years by dissolving the limestone below the soil; 2) suddenly, when the roof of a cavern collapses under its own weight

Practice Test

Multiple choice

1. c
2. c
3. d
4. c
5. b
6. b
7. a
8. d
9. b
10. d
11. e
12. a
13. b
14. c
15. b
16. c
17. e
18. e
19. d
20. b
21. c
22. a
23. b
24. d
25. a

True/false

1. T
2. F (temporary)
3. T
4. T
5. T
6. F (wide)
7. T
8. F (decreases)
9. F (rectangular)
10. T
11. T
12. F (impermeable)
13. T
14. T
15. T
16. F (gravity)
17. T
18. T
19. T
20. T
21. T
22. T
23. F (transpiration)
24. T
25. T
26. T
27. T
28. T
29. T
30. F (running water)

Written questions

1. Evaporation, primarily from the ocean, transport via the atmosphere, and eventually precipitation back to the surface. If the water falls on the continents, much of it will find its way back to the atmosphere by evaporation and transpiration. Some water, however, will be absorbed by the surface (infiltration) and some will run off back to the ocean.

2. To learn about the development of a valley, it is helpful to divide its development into three stages: youth, maturity, and old age. A youthful valley is characterized by a stream that is downcutting, waterfalls, rapids, and a V-shape. In a mature valley, the downward erosion of the stream diminishes and lateral erosion dominates. The stream is meandering and a floodplain is beginning. In old age, the floodplain is large, meandering is widespread, and the river is primarily reworking unconsolidated floodplain deposits. Natural levees, back swamps, and yazoo tributaries are also common features.

3. Porosity is the percentage of the total volume of rock or sediment that consists of pore spaces. Permeability is a measure of a material's ability to transmit water through interconnected pore spaces. It is possible for a material to have a high porosity but a low permeability if the pore spaces lack interconnections.

4. Generally, for karst topography to develop the area must be humid and underlain by soluble rock, most often limestone.

CHAPTER FOUR

Vocabulary Review

1. ice sheet
2. ephemeral stream
3. drift
4. drumlin
5. Pleistocene epoch
6. glacial striations
7. interior drainage
8. medial moraine
9. zone of wastage
10. glacial trough
11. cirque
12. dune
13. zone of accumulation
14. alluvial fan
15. steppe
16. loess
17. till
18. outwash plain
19. blowout
20. cross beds
21. glacier

Comprehensive Review

1. 1) Valley, or alpine, glaciers are relatively small streams of ice that exist in mountain areas, where they usually follow valleys originally occupied by streams. 2) Ice sheets exist on a large scale and flow out in all directions from one or more centers, completely obscuring all but the highest area of underlying terrain.
2. This upper part of the glacier consists of brittle ice that is subjected to tension when the ice flows over irregular terrain, with cracks called crevasses resulting.
3. a) Plucking occurs when a glacier flows over a fractured bedrock surface and the ice incorporates loose blocks of rock and carries them off. b) Abrasion occurs as the ice with its load of rock fragments moves along and acts as a giant rasp or file, grinding the surface below as well as the rocks within the ice.
4. a) D; b) E; c) A; d) C; e) B
5. Till is unsorted material deposited directly by the glacier. Stratified drift is characteristically sorted sediment laid down by glacial meltwater.
6. a) A desert is an arid area. b) A steppe is semiarid and is a marginal and more humid variant of the desert. It represents a transition zone that surrounds the desert and separates it from bordering humid climates.
7. B; a) C; b) D
8. a) forms at the terminus of a glacier; b) form along the sides of a valley glacier; c) a layer of till that is laid down as the glacier recedes
9. a) A; b) G; c) C; d) B
10. Periodic slides of sand down the slip face of a dune cause the slow migration of the dune in the direction of air movement.
11. The indirect effects of Ice Age glaciers included plant and animal migrations, changes in stream and river courses, adjustment of the crust by rebounding, climate changes, and the worldwide change in sea level that accompanied each advance and retreat of the ice sheets.

12. moraines, glacial striations, glacial troughs, hanging valleys, cirques, horns, artes, fiords, glacial drift, glacial erratics
13. The existence of several layers of glacial drift with well-developed zones of chemical weathering and soil formation as well as the remains of plants that require warm temperatures indicates several glacial advances separated by periods of warmer climates.
14. Wind is not capable of picking up and transporting coarse materials and, because it is not confined to channels, it can spread over large areas as well as high into the atmosphere.

Practice Test

Multiple choice

1. c	6. a	11. b	16. b	21. a
2. c	7. d	12. d	17. a	22. b
3. e	8. b	13. a	18. b	23. b
4. d	9. b	14. c	19. e	24. d
5. a	10. b	15. e	20. d	25. a

True/false

1. T	12. T
2. F (accumulation)	13. F (U-shaped)
3. T	14. F (drumlins)
4. F (lateral)	15.T
5. T	16. T
6. F (fifty)	17. T
7. T	18. F (horn)
8. T	19. T
9. F (ten)	20. F (cirques)
10. F (middle)	21. T
11. F (meters)	22. T

Written questions

1. 1) moraines; layers or ridges of till; 2) drumlins; streamlined, asymmetrical hills composed of till; 3) eskers; ridges of sand and gravel deposited by streams flowing beneath the ice, near the glacier's terminus; 4) kames; steep-sided hills formed when glacial meltwater washes sediment into openings and depressions in the stagnant, wasting terminus of a glacier (answers may vary)
2. 1) glacial trough; a U-shaped glaciated valley; 2) horn; a pyramid-shaped peak; 3) cirque; a bowl-shaped depression at the head of a valley glacier where snow accumulation and ice formation occur (answers may vary)
3. 1) plant and animal migrations; 2) changes in stream and river courses; 3) worldwide change in sea level that accompanied each advance and retreat of the ice sheets (answers may vary)
4. Periodic slides of sand down the slip face of a dune cause the slow migration of the dune in the direction of air movement.

CHAPTER FIVE

Vocabulary Review

1. continental drift
2. plate
3. sea-floor spreading
4. convergent boundary
5. paleomagnetism
6. plate tectonics
7. normal polarity
8. divergent boundary
9. hot spot
10. Pangaea
11. subduction zone
12. rift (rift valley)
13. transform boundary
14. polar wandering
15. reverse polarity
16. volcanic island arc
17. deep-ocean trench

Comprehensive Review

1. 1) fit of the continents; 2) similar fossils on different landmasses; 3) similar rock types and structures on different landmasses; 4) similar ancient climates on different landmasses

2. One of the main objections to Wegener's continental drift hypothesis was his inability to provide a mechanism that was capable of moving the continents across the globe.

3. a) C; b) A; c) B

4. The theory of plate tectonics holds that Earth's rigid outer shell consists of about twenty rigid slabs called plates that are in continuous slow motion relative to each other. These plates interact in various ways and thereby produce earthquakes, volcanoes, mountains, and the crust itself.

5. 1) oceanic-continental convergence; 2) oceanic-oceanic convergence; 3) continental-continental convergence

6. a) Paleomagnetism is used to show that the continents have moved through time and, using the patterns of magnetic reversals found on the ocean floors, to explain sea-floor spreading. b) Earthquake patterns are used to show plate boundaries and to trace the descent of slabs into the mantle. c) Ocean drilling has shown that the ocean basins are geologically youthful and confirms sea-floor spreading by the fact that the age of the deepest ocean sediment increases with increasing distance from the ridge.

7. a) D; b) B; c) C; d) F

8. 1) convection current hypothesis; large convection currents within the mantle drive plate motion; 2) slab-push and slab-pull hypothesis; as new ocean crust moves away from the ridge it cools, becomes denser, and eventually descends, pulling the trailing lithosphere along; 3) hot plumes hypothesis; hot plumes within the mantle reach the lithosphere and spread laterally, facilitating the plate motion away from the zone of upwelling.

9. a) F; b) B; c) C; d) E; e) A; f) D; g) G

Practice Test

Multiple choice

1. b	7. e	13. b	19. b	25. b
2. b	8. d	14. c	20. e	26. d
3. b	9. a	15. b	21. b	
4. e	10. b	16. a	22. b	
5. a	11. e	17. a	23. a	
6. d	12. a	18. c	24. d	

True/false

1. T	6. F (mantle)	11. F (trenches)	16. F (normal)	21. T
2. T	7. F (younger)	12. T	17. T	22. T
3. F (centimeters)	8. T	13. T	18. F (youngest)	23. F (thinnest)
4. F (divergent)	9. T	14. F (mantle)	19. T	24. T
5. T	10. T	15. T	20. T	25. T

Written questions

1. Hot spots are relatively stationary plumes of molten rock rising from Earth's mantle. According to the plate tectonics theory, as a plate moves over a hot spot, magma often penetrates the surface, thereby generating a volcanic structure. In the case of the Hawaiian Islands, as the Pacific plate moved over a hot spot, the associated igneous activity produced a chain of major volcanoes. Currently, because the rising plume of mantle material is located below it, the only active island in the chain is Hawaii. Kauai, the oldest island in the chain, formed approximately five million years ago when it was positioned over the hot spot and has now moved northwest as a result of plate motion.

2. 1) divergent boundaries—where the plates are moving apart; 2) convergent boundaries—where the plates are moving toward each other; 3) transform boundaries—where the plates are sliding past each other along faults

3. The lines of evidence used to support plate tectonics include 1) paleomagnetism (polar wandering and magnetic reversals), 2) earthquake patterns, 3) the age and distribution of ocean sediments, and 4) evidence from sea floor volcanoes that are, or were, located over hot spots.

CHAPTER SIX

Vocabulary Review

1. earthquake
2. seismic sea wave (tsunami)
3. foreshock
4. focus
5. elastic rebound
6. surface wave
7. strike-slip fault
8. epicenter
9. asthenosphere
10. fault
11. anticline
12. secondary(S) wave
13. seismogram
14. Mohorovicic discontinuity (Moho)
15. magnitude
16. inner core
17. primary(P) wave
18. lithosphere
19. syncline
20. normal fault
21. mantle
22. accretionary wedge
23. terrane
24. fault creep

Comprehensive Review

1. 1) surface waves; travel around the outer layer of Earth; 2) primary(P) waves; travel through the body of Earth with a push-pull motion in the direction the wave is traveling; 3) secondary(S) waves; travel through Earth by "shaking" the rock particles at right angles to their direction of travel
2. a) A; b) B
3. a) C; b) B; c) A
4. Primary(P) waves have a greater velocity than secondary(S) waves. P waves can travel through solids, liquids, and gasses; while S waves are only capable of being transmitted through solids.
5. By determining the difference in arrival times between the first P wave and first S wave and using a travel-time graph, three different recording stations determine how far they each are from the epicenter. Next, for each station, a circle is drawn on a globe corresponding to the station's distance from the epicenter. The point where the three circles intersect is the epicenter of the quake.
6. 1) magnitude of the earthquake; 2) proximity to a populated area
7. surrounding the Pacific Ocean basin and along the mid-ocean ridge (answers may vary); Most earthquake epicenters are closely correlated with plate boundaries.
8. a) E (including D and part of C); b) G; c) A; d) B; e) D; f) F; g) C
9. a) Denser rocks like peridotite are thought to make up the mantle and provide the lava for oceanic eruptions. b) The molten outer core is thought to be mainly iron and nickel.
10. a) hanging wall moves up relative to the footwall (letter B); b) dominant displacement is along the trend, or strike, of the fault (letter D); c) hanging wall moves down relative to the footwall (letter A); d) hanging wall moves up relative to the footwall at a low horizontal angle (letter C)
11. a) B; b) A
12. normal fault
13. a) An anticline is most commonly formed by the upfolding, or arching, of rock layers. (letter A); b) Synclines, usually found in association with anticlines, are down folds, or troughs. (letter B)

14. a) Where oceanic and continental crusts converge, the first stage in the development of a mountain belt is the formation of a subduction zone. Deposition of sediments occurs along the continental margin. Convergence of the continental block and the subducting oceanic plate leads to deformation and metamorphism of the continental margin. A volcanic island arc forms. Sediment derived from the land as well as that scraped from the subducting plate becomes plastered against the landward side of the trench forming an accretionary wedge consisting of folded, faulted, and metamorphosed sediments and volcanic debris. b) Where continental crusts converge, the continental lithosphere is too buoyant to undergo any appreciable subduction and a collision between continental fragments eventually results. The result of the collision is the formation of a mountain range.

Practice Test

Multiple choice

1. e	6. a	11. e	16. c	21. c
2. c	7. c	12. a	17. d	22. a
3. e	8. d	13. b	18. a	23. c
4. b	9. b	14. d	19. b	24. b
5. a	10. e	15. c	20. c	25. c

True/false

1. T	8. F (vertical)	14. T	20. T
2. F (aftershocks)	9. F (seismology)	15. F (above)	21. T
3. F (two)	10. T	16. T	22. F (less)
4. T	11. F (can)	17. T	23. T
5. T	12. F (before)	18. T	24. F (convergent)
6. T	13. F (Mohorovicic)	19. T	25. T
7. T			

Written questions

1. An earthquake is the vibration of Earth produced by the rapid release of energy, usually along a fault. This energy radiates in all directions from the earthquake's source, called the focus, in the form of waves.

2. Oceanic crust is basaltic in composition; while the continental crust has an average composition similar to that of granite. The composition of the mantle is thought to be similar to the rock peridotite, which contains iron and magnesium-rich silicate minerals. Both the inner and outer cores are thought to be enriched in iron and nickel, with lesser amounts of other heavy elements.

3. Most of Earth's major mountain systems have formed along convergent plate boundaries. Convergence can occur between one oceanic and one continental plate, between two oceanic plates, or between continental plates. In the first two, subduction zones often develop and the sediments and volcanics are squeezed, faulted, and deformed into mountain ranges as the plates converge. When continents collide, a collision between continental fragments deforms the continental material into mountains.

CHAPTER SEVEN

Vocabulary Review

1. pyroclasts
2. caldera
3. aa flow
4. volcanic neck
5. dike
6. vent
7. cinder cone
8. batholith
9. shield volcano
10. flood basalt
11. pahoehoe flow
12. hot spot
13. composite cone
14. volcano
15. laccolith
16. fissure
17. sill
18. nue ardente
19. viscosity
20. mantle plumes

Comprehensive Review

1. 1) the magma's composition; 2) the magma's temperature; 3) the amount of dissolved gases in the magma
2. a) B (Mauna Loa); Shield volcanoes have recurring eruptions over a long span of time and are composed of large amounts of basaltic lava. b) A (Paricutin); The eruption history of a cinder cone is short and produces a relatively small cone composed of loose pyroclastic material. c) C (Fujiyama); A composite cone may extrude viscous lava for a long period and then, suddenly, the eruptive style changes and the volcano violently ejects pyroclastic material. Composite cones represent the most violent type of volcanic activity.
3. water vapor, carbon dioxide, nitrogen, sulfur gases
4. a) the igneous body cuts across existing sedimentary beds; b) the igneous body is parallel to existing sedimentary beds
5. The viscosity of magma is affected by its temperature, silica content (the more silica, the greater the viscosity), and quantity of dissolved gases.
6. a) C; b) E; c) F; d) D
7. When compared to granitic magma, basaltic magma contains less silica, and is less viscous.
8. 1) Along the oceanic ridge system, as the rigid lithosphere pulls apart, pressure is lessened and the partial melting of the underlying mantle rocks produces large quantities of basaltic magma. 2) Adjacent to ocean trenches magma is generated by the partial melting of descending slabs of oceanic crust. The melt slowly rises upward toward the surface because it is less dense than the surrounding rock. 3) Intraplate volcanism may be the result of rising plumes of hot mantle material which produce hot spots, volcanic regions a few hundred kilometers across.

Practice Test

Multiple choice

1. b	6. e	11. b	16. e	21. d
2. b	7. a	12. b	17. a	22. c
3. b	8. e	13. a	18. a	23. a
4. c	9. d	14. e	19. b	
5. b	10. c	15. a	20. d	

True/false

1. F (convergent)	6. T	11. F (silica)	16. F (sills)
2. T	7. F (less)	12. F (composite)	17. T
3. T	8. F (crater)	13. T	18. T
4. T	9. T	14. T	
5. F (cinder)	10. F (increase)	15. T	

Written questions

1. Magma and lava both refer to the material from which igneous rocks form. However, magma is molten rock below Earth's surface, including dissolved gases and crystals, while lava refers to molten rock that reaches Earth's surface.

2. 1) Shield cones are among the largest on Earth. These gently sloping domes are associated with relatively quiet eruptions of fluid basaltic lava. 2) Cinder cones are composed almost exclusively of pyroclastic material, are steep-sided, and are the smallest of the volcanoes. 3) Composite cones, as the name suggests, are composed of alternating layers of lava and pyroclastic debris. Their slopes are steeper than those of a shield volcano, but more gentle than those of a cinder cone.

3. 1) Dikes are tabular, discordant plutons. 2) Sills are tabular, concordant plutons. 3) Batholiths, by far the largest of all intrusive features, are massive, discordant plutons. (answers may vary)

CHAPTER EIGHT

Vocabulary Review

1. uniformitarianism
2. Cenozoic era
3. absolute date
4. correlation
5. half-life
6. catastrophism
7. relative dating
8. Precambrian
9. original horizontality
10. law of superposition
11. unconformity
12. fossil
13. radioactivity
14. period
15. inclusions
16. Paleozoic era
17. index fossil
18. principle of fossil succession
19. radiometric dating
20. eon

Comprehensive Review

1. Catastrophism states that Earth's landscape has developed by great catastrophes over a short span of time. On the other hand, uniformitarianism states that the physical, chemical, and biological laws that operate today have also operated in the geologic past. Uniformitarianism implies that the forces and processes that we observe today have been at work for a very long time.

2. a) Large coal swamps flourished in North America before the extinction of dinosaurs. b) Dinosaurs became extinct about 66 million years ago.

3. a) In an undeformed sequence of sedimentary rocks, each bed is older than the one above it and younger than the one below. b) Most layers of sediment are deposited in a horizontal position. c) Fossil organisms succeed each other in a definite and determinable order, and therefore any time period can be recognized by its fossil content.

4. a) A; b) C; c) B

5. 1) rapid burial; 2) possession of hardparts

6. a) a; b) c

7. An atom's atomic number is the number of protons in the nucleus, while the mass number is determined by adding together the number of protons and neutrons in the nucleus.

8. 1) emission of alpha particles (2 protons and 2 neutrons) from the nucleus; 2) emission of beta particles, or electrons (derived from neutrons), from the nucleus; 3) electrons are captured by the nucleus and combine with protons to form an additional neutron

9. Because the rates of decay for many isotopes have been precisely measured and do not vary under the physical conditions that exist in Earth's outer layers.

10. a) The Cenozoic era is often described as the era of "recent life" or the "age of mammals," including human development. b) During the Mesozoic ("middle life") era, dinosaurs are dominant and the first birds and flowering plants evolve. c) Among the events of Paleozoic ("ancient life") era are the first organisms with shells, the first fishes, first land plants, amphibians abundant, large coal swamps, and the first reptiles.

11. The primary problem is assigning absolute dates to units of time of the geologic time scale is the fact that not all rocks can be dated radiometrically.
12. a) younger (principle of cross-cutting relationships); b) younger (law of superposition); c) younger; d) older; e) older
13. 100,000 years (two half-lives, each 50,000 years)

Practice Test

Multiple choice

1. a	6. c	11. b	16. b	21. b
2. b	7. c	12. b	17. b	22. d
3. d	8. a	13. c	18. a	23. d
4. e	9. a	14. b	19. a	24. e
5. e	10. d	15. c	20. e	25. b

True/false

1. F (relative)	6. T	11. F (increases)	16. T
2. F (long)	7. F (cannot)	12. T	17. T
3. T	8. F (protons)	13. T	18. F (uniformitarianism)
4. T	9. T	14. T	19. T
5. F (hard)	10. F (younger)	15. T	20. T

Written questions

1. Radiometric dating uses the radioactive decay of particular isotopes to arrive at an absolute date, which pinpoints the time in history when something took place. For example, the extinction of the dinosaurs about 66 million years ago. Relative dating involves placing rocks or events in their proper sequence using various laws and principles. Relative dating cannot tell us how long ago something took place, only that it followed one event and preceded another.
2. 1) law of superposition;) principle of original horizontality; 3) principle of cross-cutting relationships
3. Fossils help researchers understand past environmental conditions, are useful as time indicators, and play a key role in correlating rocks of similar ages that are from different places.

CHAPTER NINE

Vocabulary Review

1. oceanography
2. salinity
3. deep-ocean trench
4. continental shelf
5. turbidity current
6. abyssal plain
7. mid-ocean ridge
8. outgassing
9. terrigenous sediment
10. atoll
11. halocline
12. coral reef
13. seamount
14. hydrogenous sediment
15. thermocline
16. biogenous sediment
17. continental rise
18. guyot
19. rift zone
20. continental slope
21. echo sounder

Comprehensive Review

1. In the Northern Hemisphere nearly 61 percent of the surface is water. About 81 percent of the Southern Hemisphere is water. The Northern Hemisphere has about twice as much land (39 percent compared to 19 percent of the surface) as the Southern Hemisphere.
2. In the low and middle latitudes, both temperature and salinity are often highest at the surface, decrease rapidly below the surface zone, and then fall off slowly to the ocean floor.
3. chlorine and sodium
4. a) D; b) C; c) G; d) F; e) B
5. Abyssal plains are incredibly flat features found adjacent to continental slopes on ocean basin floors in all oceans. They consist of thick accumulations of sediment transported by turbidity currents far out to sea. (letter E)
6. 1) weathering of rocks on the continents; 2) from Earth's interior through volcanic eruptions, a process called outgassing.
7. In the dry subtropics, where evaporation is high, the ocean's surface salinity is high because water is removed by evaporation, leaving the salts behind. Conversely, in the equatorial regions heavy precipitation dilutes ocean waters and lower salinities prevail.
8. Submarine canyons that are not seaward extensions of river valleys have probably been excavated by turbidity currents as these sand and mud laden currents repeatedly sweep downslope.
9. 1) warm waters with and average annual temperature about 24°C (75°F); 2) clear sunlit water; 3) a depth of water no greater than 45 meters (150 feet)
10. Coral islands, called atolls, form on the flanks of sinking volcanic islands from corals that continue to build the reef complex upward.
11. 1) Terrigenous sediment consists primarily of mineral grains that were weathered from continental rocks and transported to the ocean. 2) Biogenous sediment consists of shells and skeletons of marine animals and plants. 3) Hydrogenous sediment consists of minerals that crystallize directly from seawater through various chemical reactions.
12. a) C; b) F; c) B; d) A; e) E

Practice Test

Multiple choice

1. c	8. a	15. c	22. d
2. a	9. c	16. b	23. b
3. c	10. c	17. d	24. b
4. b	11. b	18. a	25. d
5. c	12. c	19. e	26. b
6. a	13. a	20. c	27. c
7. e	14. b	21. e	28. c

True/false

1. T	5. F (slope)	9. F (mud)	13. T
2. F (Atlantic)	6. T	10. T	14. T
3. T	7. T	11. T	15. T
4. T	8. F (lower)	12. T	16. F (slow)

Written questions

1. Salinity refers to the proportion of dissolved salts to pure water in the ocean, generally expressed in parts-per-thousand (‰). Salinity variations in the open ocean normally range from 33‰to 37‰.
2. 1) continental margins; 2) ocean basin floor; 3) mid-ocean ridges
3. Mid-ocean ridges are spreading centers; that is, they are divergent plate boundaries where magma from Earth's interior rises and forms new oceanic crust. Ocean trenches form where lithospheric plates plunge into the mantle. Thus, these long, narrow, linear depressions are associated with convergent plate boundaries and the destruction of crustal material.

CHAPTER TEN

Vocabulary Review

1. spit
2. beach drift
3. gyre
4. wave height
5. thermohaline circulation
6. surf
7. Coriolis effect
8. wave of oscillation
9. wavelength
10. barrier island
11. tidal current
12. wave of translation
13. upwelling
14. tombolo
15. wave period
16. sea stack
17. emergent coast
18. estuary
19. tidal flat
20. wave refraction
21. fetch
22. longshore current
23. wave-cut platform
24. tide
25. groin

Comprehensive Review

1. The Coriolis effect, the deflective force of Earth's rotation on all free-moving objects, causes the movement of ocean waters to be deflected to the right, forming clockwise gyres, in the Northern Hemisphere and to the left, forming counterclockwise gyres, in the Southern Hemisphere.
2. a) C; b) F; c) D; d) H; e) G; f) A; g) B; h) E
3. In addition to influencing temperatures of adjacent land areas, cold currents also transform some tropical deserts into relatively cool, damp places that are often shrouded in fog.
4. 1) the shape of the coastline; 2) the configuration of the ocean basin
5. a) B; b) A
6. a) crest; b) trough; c) wavelength; d) wave height
7. 1) wind speed; 2) length of time the wind has blown; 3) the distance that the wind has traveled across the open water, called fetch
8. In a wave of oscillation the wave form or shape moves forward, not the water itself. A wave of translation is the turbulent advance of water created by breaking waves.
9. 1) temperature; 2) salinity
10. 1) Arctic waters; 2) Antarctic waters
11. Wave refraction in a bay causes waves to diverge and expend less energy. As a consequence of wave refraction, sediment often accumulates and forms sheltered sandy beaches.
12. Sediment is transported along a coast by 1) beach drift, which transports sediment in a zigzag pattern along the beach and 2) by longshore currents, where turbulent water moves sediment in the surf zone parallel to the shore.
13. a) F; b) B; c) C; d) A; e) E; f) D

14. 1) as spits that were subsequently severed from the mainland by wave erosion; 2) as former sand dune ridges that originated along the shore during the last glacial period, when sea level was lower
15. 1) the topography and composition of the land; 2) prevailing winds and weather patterns; 3) the configuration of the coastline and nearshore areas (answers may vary)
16. a) Emergent coasts develop either because an area experiences uplift or as a result of a drop in sea level. Feature: elevated wave-cut platform; b) Submergent coasts are created when sea level rises or the land adjacent to the sea subsides. Feature: estuary

Practice Test

Multiple choice

1. b	6. b	11. b	16. a	21. a
2. b	7. b	12. c	17. c	22. a
3. d	8. b	13. e	18. d	23. c
4. a	9. d	14. c	19. c	24. c
5. e	10. b	15. d	20. a	25. b

True/false

1. T	4. T	7. T	10. T	13. F (right)
2. F (more)	5. T	8. T	11. T	14. F (surf)
3. F (fetch)	6. T	9. F (arch)	12. T	15. T

Written questions

1. Coastal upwelling occurs where winds are blowing equatorward and parallel to the coast. Due to the Coriolis effect, the surface water moves away from the shore area and is replaced by cold water with greater concentrations of dissolved nutrients. The nutrient enriched waters from below promote the growth of plankton, which in turn supports extensive populations of fish.

2. Deep-ocean circulation is called thermohaline circulation. It occurs when water at the surface is made colder and/or more salty and becomes more dense. The cold, dense water sinks toward the ocean bottom and flows away from its source, which is generally near the poles.

3. Tides result from the gravitational force exerted on Earth by the moon, and to a lesser extent, the sun. The gravitational force causes water bulges on both the side of Earth toward the moon and the side away from it. These tidal bulges remain in place while Earth rotates "through" them, producing periods of high and low water at any one location along a coast.

4. In a wave of oscillation, each water particle moves in a nearly circular path during the passage of the wave. As a wave approaches the shore, the movement of water particles at the base of the wave are interfered with by the bottom. As the speed and length of the wave diminish, wave height increases. Eventually, water particles at the top of the wave pitch forward and the wave collapses, or breaks, and forms surf. What had been a wave of oscillation now becomes a wave of translation in which the water advances up the shore.

CHAPTER ELEVEN

Vocabulary Review

1. troposphere
2. Weather
3. Rotation
4. thermosphere
5. Climate
6. isotherm
7. Tropic of Cancer
8. element
9. Convection
10. inclination of the axis
11. equinox
12. environmental lapse rate
13. Tropic of Capricorn
14. circle of illumination
15. mesosphere
16. radiation (electromagnetic radiation)
17. revolution
18. Conduction
19. stratosphere
20. infrared
21. solstice
22. ultraviolet (UV)
23. albedo
24. greenhouse effect

Comprehensive Review

1. Weather is the state of the atmosphere at a particular place for a short period of time. Climate, on the other hand, is a generalization of the weather conditions of a place over a long period of time.
2. a) C; b) F; c) E; d) G; e) D; f) A
3. 1) rotation: the spinning of Earth about its axis; 2) revolution: the movement of Earth in its orbit around the sun
4. 1) air temperature; 2) humidity; 3) type and amount of cloudiness; 4) type and amount of precipitation; 5) air pressure; 6) the speed and direction of the wind
5. 1) Dust particles act as surfaces upon which water vapor may condense. 2) Dust may absorb or reflect incoming solar radiation.
6. 1) When the sun is high in the sky, the solar rays are most concentrated. 2) The angle of the sun determines the amount of atmosphere the rays must penetrate.
7. a) spring equinox, March 21-22, equator; b) summer solstice, June 21-22, 23°N (Tropic of Cancer); c) autumnal equinox, September 22-23, equator; d) winter solstice, December 21-22, 23°S (Tropic of Capricorn)
8. nitrogen (78%) and oxygen (21%)
9. Ozone has three oxygen atoms per molecule, while oxygen has two atoms per molecule. Ozone in the atmosphere absorbs the potentially harmful ultraviolet (UV) radiation from the sun.
10. a) 20; b) 50; c) 22
11. Most of the energy that is absorbed at Earth's surface is reradiated skyward in the form of longer wavelengths than solar radiation. This terrestrial radiation is readily absorbed by water vapor and carbon dioxide in the atmosphere, thereby heating the atmosphere from the ground up rather than vice versa.

12. carbon dioxide and water vapor
13. 1) Conduction is the transfer of heat through matter by molecular activity. 2) Convection is the transfer of heat by the movement of a mass or substance from one place to another. 3) Radiation is the transfer of energy (heat) through space by electromagnetic waves. It is the only mechanism of heat transfer that can transmit heat through the relative emptiness of space.
14. 1) Shifts in temperature and rainfall patterns that could disrupt agriculture worldwide. 2) A gradual rise in sea level due to a warmer ocean and melting of glaciers that would accelerate shoreline erosion in many coastal areas. 3) Changing storm tracks and a higher frequency and greater intensity of hurricanes. (answers may vary)
15. a) The daily mean temperature is determined by adding the maximum and minimum temperatures and then dividing by two. b) The annual mean temperature is an average of the twelve monthly means. c) The annual temperature range is computed by finding the difference between the highest and lowest monthly means.
16. a) Land heats more rapidly and to higher temperatures than water. Land also cools more rapidly and to lower temperatures than water. b) Due to the mechanism by which the atmosphere is heated, temperature generally decreases with an increase in altitude. c) A windward coastal location will often experience cool summers and mild winters when compared to an inland station at the same latitude. Leeward coastal sites will tend to have a more continental temperature pattern because the winds do not carry the ocean's influence onshore.
17. January and July are selected most often for analysis because, for most locations, they represent the temperature extremes.

Practice Test

Multiple choice

1. a	7. c	13. e	19. c	25. d
2. d	8. a	14. b	20. d	26. e
3. a	9. b	15. c	21. b	27. e
4. b	10. b	16. c	22. b	28. c
5. d	11. a	17. b	23. a	29. b
6. b	12. c	18. a	24. b	30. c

True/false

1. T	9. T	17. T
2. F (weather)	10. F (temperature)	18. T
3. F (lower)	11. T	19. F (faster)
4. T	12. F (troposphere)	20. T
5. T	13. F (illumination)	21. T
6. F (conduction)	14. F (winter)	22. T
7. T	15. F (gradual)	23. F (smaller)
8. T	16. T	

Written questions

1. CFC's, chlorofluorocarbons, are a group of chemicals that were once commonly used as propellants for aerosol sprays and in the production of certain plastics and refrigerants. When CFC's reach the stratosphere, chlorine breaks up some of the ozone molecules. The net effect is depletion of the ozone layer and a reduction in the layer's ability to absorb harmful ultraviolet (UV) radiation.

2. The amount of solar energy reaching places on Earth's surface varies with the seasons because Earth's orientation to the sun continually changes as it travels along its orbit. This changing orientation is due to the facts that 1) Earth's axis is inclined and 2) the axis remains pointed in the same direction (toward the North Star) as Earth journeys around the sun.

3. The atmosphere allows most of the short-wave solar energy to pass through and be absorbed at Earth's surface. Earth reradiates this energy in the form of longer wavelength terrestrial radiation, which is absorbed by carbon dioxide and water vapor in the lower atmosphere. This mechanism, referred to as the greenhouse effect, explains the general drop in temperature with increasing altitude experienced in the troposphere.

CHAPTER TWELVE

Vocabulary Review

1. humidity
2. condensation
3. Relative humidity
4. Orographic lifting
5. vapor pressure
6. adiabatic temperature change
7. Latent heat
8. cloud
9. sleet
10. saturation
11. wet adiabatic rate
12. Sublimation
13. stratus
14. specific humidity
15. Frontal wedging
16. condensation nuclei
17. melting
18. rain
19. dew point
20. Rime
21. deposition
22. hail
23. dry adiabatic rate
24. hygroscopic nuclei
25. evaporation
26. Radiation fog
27. fog
28. rainshadow desert
29. hygrometer
30. convergence

Comprehensive Review

1. a) D; b) C; c) F; d) B; e) A; f) E
2. 1) adding moisture to the air; 2) lowering the air temperature
3. a) 25% (5 grams per kilogram/20 grams per kilogram); b) 60% (3 grams per kilogram/5 grams per kilogram)
4. a) 15°C (59°F); b) 14°F (-10°C)
5. a) The psychrometer consists of two thermometers mounted side by side: the dry-bulb, which gives the present temperature, and the wet-bulb, which has a moist, thin muslin wick tied around the end. The lower the relative humidity, the more evaporation (and hence cooling) there will be from the wet-bulb thermometer. Thus, the greater the difference between the wet and dry bulb temperatures, the lower the relative humidity. b) With a hair hygrometer, the length of the hair or other synthetic fibers changes in proportion to changes in the relative humidity. They lengthen as relative humidity increases and shrink as the relative humidity drops.
6. 42%
7. Once the air is cooled sufficiently to reach the dew point and condensation occurs, latent heat of condensation stored in the water vapor will be liberated, thereby reducing the rate at which the air cools.
8. Stable air resists vertical movement because the environmental lapse rate is less than the wet adiabatic rate. Unstable air wants to rise because it is less dense than the surrounding air. Unstable conditions prevail when the environmental lapse rate is greater than the dry adiabatic rate.
9. a) Orographic lifting occurs when elevated terrains, such as mountain barriers, act as barriers to flowing air. b) Frontal wedging occurs when cool air acts as a barrier over which warmer, less dense air rises. c) Convergence occurs whenever air masses flow together. Because the air must go somewhere, it moves upward.

10. 1) the form of the cloud; 2) the height of the cloud
11. a) B; b) A
12. a) Steam fog, a type of evaporation fog, occurs when cool air moves over warm water and moisture evaporated from the water surface condenses. b) Frontal fog occurs when warm air is lifted over colder air by frontal wedging. If the resulting clouds yield rain, and the cold air below is near the dew point, enough rain will evaporate to produce fog.
13. 1) The ice crystal process occurs when ice crystals in a cloud collect available water vapor, eventually growing large enough to fall as snowflakes. When the surface temperature is about 4°C (39°F) or higher, snowflakes usually melt before they reach the ground and continue their descent as rain. 2) The collision-coalescence process occurs when large droplets form on large hygroscopic condensation nuclei and, because the bigger droplets fall faster in a cloud, they collide and join with smaller water droplets. After many collisions the droplets are large enough to fall to the ground as rain.
14. a) Sleet, a wintertime phenomenon, forms when raindrops freeze as they fall through colder air that lies below warmer air. b) Hail is produced in a large cumulonimbus cloud with violent updrafts and an abundance of supercooled water. As they fall through the cloud, ice pellets grow by collecting supercooled water droplets. The process continues until the hailstone encounters a downdraft or grows too heavy to remain suspended by the thunderstorm's updraft.

Practice Test

Multiple choice

1. d	7. b	13. b	19. b	25. a
2. b	8. e	14. a	20. d	26. b
3. b	9. c	15. c	21. a	27. d
4. a	10. b	16. b	22. d	28. c
5. b	11. a	17. d	23. a	29. c
6. c	12. c	18. e	24. a	30. a

True/false

1. T	7. T	13. F (decrease)
2. F (calorie)	8. T	14. T
3. F (height)	9. F (wet-bulb)	15. F (vaporization)
4. T	10. T	16. T
5. F (cool)	11. T	17. F (fog)
6. T	12. F (wet)	18. T

Written questions

1. Clouds are classified on the basis of their form (cirrus, cumulus, and stratus) and height (high, middle, low, and clouds of vertical development).
2. As air rises, it expands because air pressure decreases with an increase in altitude. Expanding air cools because it pushes (does work on) on the surrounding air and expends energy.
3. Stable air resists vertical movement because the environmental lapse rate is less than the wet adiabatic rate. Unstable air wants to rise because it is less dense than the surrounding air. Unstable conditions prevail when the environmental lapse rate is greater than the dry adiabatic rate.
4. Answers will vary. Depending on the place and season, the process will be either orographic lifting, frontal wedging, or convergence. In either case, when air rises it will expand and cool adiabatically. If sufficient cooling takes place, the dew point may be reached and condensation may result in the formation of clouds and perhaps precipitation.

CHAPTER THIRTEEN

Vocabulary Review

1. polar front
2. isobar
3. monsoon
4. Coriolis effect
5. wind
6. Barometric tendency
7. aneroid barometer
8. prevailing wind
9. wind vane
10. cyclone
11. Convergence
12. land breeze
13. equatorial low
14. chinook
15. sea breeze
16. cup anemometer
17. pressure gradient
18. jet stream
19. anticyclone
20. subpolar low
21. polar easterlies
22. barograph
23. divergence
24. subtropical high
25. westerlies
26. trade winds

Comprehensive Review

1. a) The mercury barometer is an instrument used to measure air pressure. It consists of glass tube, closed at one end, filled with mercury. The tube is inverted in a reservoir (pan) of mercury. If the air pressure decreases, the height of the column of mercury in the tube falls; if pressure increases, the column rises. b) The wind vane is used to determine wind direction by pointing into the wind. c) The aneroid barometer measures air pressure using partially evacuated metal chambers that are compressed as air pressure increases and expand as pressure decreases. d) The cup anemometer measures wind speed using several cups mounted on a shaft that is rotated by the moving wind. e) Barographs are aneroid barometers that continuously record pressure changes.
2. 1) pressure gradient force: the amount of pressure change over a given distance; 2) Coriolis effect: the deflective force of Earth's rotation on all free-moving objects. Deflection is to the right in the Northern Hemisphere. 3) friction with Earth's surface: slows down air movement within the first few kilometers of Earth's surface and, as a consequence, alters wind direction
3. (The diagrams in Figure 13.1 should show the following pressures and surface air movements.) Northern Hemisphere high: highest pressure in center, air moves out (divergence), clockwise; Southern Hemisphere high: highest pressure in center, air moves out (divergence), counterclockwise; Northern Hemisphere low: lowest pressure in center, air moves inward (convergence), counterclockwise; Southern Hemisphere low: lowest pressure in center, air moves inward (convergence), clockwise
4. a) inward and counterclockwise; b) outward and counterclockwise
5. a) B; b) A; c) B; d) A; e) B
6. In a high pressure system (anticyclone) outflow near the surface is accompanied by convergence aloft and general subsidence of the air column. Because descending air is compressed and warmed, cloud formation and precipitation are unlikely and "fair" weather can usually be expected.

7. A: subpolar low; B: subtropical high; C: equatorial low; D: subtropical high; E: subpolar low; F: polar easterlies; G: westerlies; H: trade winds; I: trade winds; J: westerlies; K: polar easterlies

8. a) subpolar low; b) subtropical high; c) equatorial low; d) polar high

9. a) warm, dry winds often found on the leeward sides of mountains, especially the eastern slopes of the Rockies; b) cooler air over the water (higher pressure) moves toward the warmer land (lower pressure)

10. a) northeast; b) west; c) south

Practice Test

Multiple choice

1. d	6. c	11. c	16. e	21. b
2. a	7. a	12. a	17. d	22. b
3. a	8. d	13. e	18. b	23. a
4. c	9. d	14. d	19. c	24. b
5. b	10. c	15. e	20. d	25. e

True/false

1. T	5. T	9. F (dry)	13. F (subtropical)
2. F (faster)	6. T	10. T	14. T
3. T	7. T	11. T	15. F (convergence)
4. F (right)	8. F (cyclone)	12. F (greater)	16. F (colder)

Written questions

1. The Coriolis effect, the deflective force of Earth's rotation on all free-moving objects, causes air to be deflected to the right of its path of motion in the Northern Hemisphere and to the left in the Southern Hemisphere.

2. A drop in barometric pressure indicates the approach of a low pressure system, called a cyclone. The cyclone is associated with converging surface winds and ascending air; hence, adiabatic cooling, cloud formation, and precipitation often occur. Conversely, an anticyclone, with its high air pressure, is associated with subsidence of the air column and divergence at the surface. Because descending air is compressed and warmed, cloud formation and precipitation are unlikely in an anticyclone and "fair" weather can usually be expected.

3. In a Northern Hemisphere cyclone (low pressure), surface winds are inward (convergence) and counterclockwise. As a consequence of surface convergence, there is a slow, net upward movement of air near the center of the cyclone. Furthermore, the surface convergence found in a cyclone is balanced by divergence of the air aloft.

CHAPTER FOURTEEN

Vocabulary Review

1. air mass
2. hurricane
3. front
4. tropical storm
5. source region
6. continental (c) air mass
7. warm front
8. middle-latitude cyclone
9. eye wall
10. tropical (T) air mass
11. tropical depression
12. eye
13. air-mass weather
14. cold front
15. tornado
16. maritime (m) air mass
17. tornado watch
18. Doppler radar
19. thunderstorm
20. storm surge
21. tornado warning
22. polar (P) air mass

Comprehensive Review

1. Air masses are classified according to their source region using two criteria. 1) The nature of the surface (land or water) where the air mass originates is identified using a lower case letter such as c (continental) or m (maritime). 2) The latitude where the air mass originates is identified using an upper case letter such as P (polar) or T (tropical). Therefore, a cP air mass would have a continental polar source region.

2. A: maritime polar (mP)—cool, moist; B: continental polar (cP)—cold and dry in winter; F: continental tropical (cT)—hot and dry in summer; G: maritime tropical (mT)—warm, moist

3. continental polar (cP) from central Canada and maritime tropical (mT) from the Gulf of Mexico

4. (See textbook Figures 14.3 and 14.5) 1) A warm front occurs where the surface position of a front moves so that warm air occupies territory formerly covered by cooler air. The boundary separating the cooler air from the warmer air above has a small slope (about 1:200). 2) A cold front occurs when cold air actively advances into a region occupied by warmer air. On the average, the boundary separating the cold and warm air is about twice as steep as the slope along a warm front.

5. a) warm front; b) cold front; c) maritime tropical (mT); d) continental polar (cP); e) southwest to south; f) northwest to north; g) thunderstorms and precipitation with the passage of the cold front, followed by cooler, drier conditions as the cP air moves over the area, change in wind direction from southwest to northwest after the cold front passes

6. a) surface: inward (convergent), counterclockwise (Northern Hemisphere), rising near center; aloft: divergence; b) surface: outward (divergence). clockwise (Northern Hemisphere), subsiding near center; aloft: convergence

7. 1) rate of movement; 2) steepness of slope

8. a) All thunderstorms require warm, moist air, which, when lifted, will release sufficient latent heat to provide the buoyancy necessary to maintain its upward flight. b) Tornadoes form in association with severe thunderstorms spawned along the cold front of a middle-latitude cyclone. These violent windstorms that take the form of a rapidly rotating column of air that extends downward from a cumulonimbus cloud are products of the interaction between strong updrafts in the thunderstorm and winds in the troposphere.

9. 1) when they move over waters that cannot supply warm, moist, tropical air; 2) when they move onto land; 3) when they reach a location where the large-scale flow aloft is unfavorable
10. 1) wind damage; 2) storm surge damage; 3) inland freshwater flooding

Practice Test

Multiple choice

1. b	6. a	11. b	16. b	21. c
2. b	7. d	12. e	17. b	22. c
3. b	8. b	13. c	18. a	23. e
4. a	9. d	14. b	19. d	24. d
5. b	10. d	15. a	20. b	25. d

True/false

1. T	6. F (eastward)	11. T	16. F (central)	20. T
2. F (one)	7. F (maritime)	12. F (cold)	17. T	21. T
3. T	8. T	13. T (twice)	18. T	22. T
4. T	9. T	14. T	19. T	
5. F (more)	10. T	15. T		

Written questions

1. A continental polar (cP) air mass that originates in central Canada will most likely be cold and dry in the winter and cool and dry in the summer. A maritime tropical air mass with its source region in the Gulf of Mexico will be warm and moist.

2. As the warm front approaches, winds would blow from the east or southeast and air pressure would drop steadily. Cirrus clouds would be sighted first, followed by progressively lower clouds, and perhaps nimbostratus clouds. Gentle precipitation could occur as the nimbostratus clouds moved overhead. As the warm front passed, temperatures would rise, precipitation would cease, and winds would shift to the south or southwest. Further, the sky clears and the pressure tendency steadies.

Later, with the approach of the cold front, cumulonimbus clouds fill much of the sky and bring the likelihood of heavy precipitation and a possibility of hail and tornado activity. The passage of the cold front is accompanied by a drop in temperature, clearing sky, a wind shift to the northwest or north, and rising air pressure. Fair weather can probably be expected for the next few days.

3. A tornado watch alerts the public to the fact that conditions are right for the formation of tornadoes; whereas a tornado warning is issued when a tornado has actually been sighted in an area or is indicated by radar.

CHAPTER FIFTEEN

Vocabulary Review

1. retrograde motion
2. terrestrial planet
3. asteroid
4. maria
5. geocentric
6. Jovian planet
7. astronomical unit
8. Lunar regolith
9. meteor
10. escape velocity
11. stony meteorite
12. comet
13. celestial sphere
14. meteorite
15. meteor shower
16. Ptolemaic system
17. Cassini division
18. iron meteorite
19. nebular hypothesis
20. stony-iron meteorite

Comprehensive Review

1. The geocentric model held by the early Greeks proposed that Earth was a sphere that stayed motionless at the center of the universe. Orbiting Earth were the moon, sun, and the known planets, Mercury through Jupiter. Beyond the planets was a transparent, hollow sphere, called the celestial sphere, on which the stars traveled daily around Earth.

2. Retrograde motion is the apparent periodic westward drift of a planet in the sky that results from a combination of the motion of Earth and the planet's own motion around the sun. To explain retrograde motion, the geocentric model of Ptolemy had the planets moving in small circles, called epicycles, as they revolved around Earth along large circles, called deferents.

3. Early Greeks rejected the idea of a rotating Earth because Earth exhibited no sense of motion and seemed too large to be movable.

4. a) Galileo Galilei; b) Nicolaus Copernicus; c) Tycho Brahe; d) Johannes Kepler; e) Nicolaus Copernicus; f) Johannes Kepler; g) Galileo Galilei; h) Galileo Galilei; i) Sir Isaac Newton

5. Stellar parallax is the apparent shift in the position of a nearby star, when observed from extreme points in Earth's orbit six months apart, with respect to the more distant stars.

6. 1) The path of each planet around the sun is an ellipse with the sun at one focus. 2) Each planet revolves so that an imaginary line connecting it to the sun sweeps over equal areas in equal intervals of time. 3) The orbital periods of the planets and their distances to the sun are proportional.

7. 1) four moons orbiting Jupiter; 2) the planets are circular disks rather than just points of light; 3) Venus has phases just like the moon; 4) the moon's surface is not a smooth glass sphere

8. a) The terrestrial planets (Mercury, Venus, Earth, and Mars) are all similar to Earth in their physical characteristics. b) The Jovian planets (Jupiter, Saturn, Uranus, and Neptune) are all Jupiterlike in their physical characteristics.

9. a) The Jovian planets are all large compared to the terrestrial planets. b) Because the Jovian planets contain a large percentage of gases, their densities are less than those of the terrestrial planets. c) The Jovian planets have shorter periods of rotation. d) The Jovian planets are all more massive than the terrestrial planets. e) The Jovian planets have more moons. f) The periods of revolution of the Jovian planets are longer than those of the terrestrial planets. g) The terrestrial planets are mostly rocky and metallic material, with minor amounts of gases. The Jovian planets, on the other hand, contain a large percentage of gases (hydrogen and helium), with varying amounts of ices (mostly water, ammonia, and methane).

10. a) hydrogen and helium; b) principally silicate minerals and metallic iron; c) ammonia, methane, carbon dioxide, and water
11. The nebular hypothesis suggests that all bodies of the solar system began forming about 5 billion years ago from an enormous nebular cloud consisting of approximately 80 percent hydrogen, 15 percent helium, and a few percent of all other heavier elements known to exist. As the cloud collapsed, it began to rotate and assume a disk shape. Inside, relatively small contractions formed the nuclei from which the planets would develop. The greatest concentration of material was pulled toward the center, eventually forming the protosun.
12. a) A; b) B; c) D; d) C; e) C
13. Because the gravitational attraction at the lunar surface is only one-sixth of that experienced on Earth's surface.
14. Lunar maria basins are enormous impact craters that were flooded with layer upon layer of very fluid basaltic lava.
15. a) Mars; b) Venus; c) Jupiter; d) Saturn; e) Mercury; f) Saturn; g) Venus; h) Uranus; i) Jupiter; j) Neptune; k) Saturn; l) Mars; m) Jupiter; n) Venus; o) Neptune; p) Mercury; q) Pluto; r) Mars.
16. 1) they might have formed from the breakup of a planet; 2) perhaps several large bodies once coexisted in close proximity and their collisions produced numerous smaller objects
17. a) B; b) A; c) C; d) D
18. radiation pressure: pushes dust particles away from the coma; 2) solar wind: responsible for moving the ionized gases
19. A meteor is a brilliant streak of light produced when a meteoroid enters Earth's atmosphere and burns up. A meteorite is any portion of a meteoroid that survives its traverse through Earth's atmosphere and strikes the surface.
20. Meteorites are classified by their composition: 1) irons: mostly iron with 5-10 percent nickel; 2) stony: silicate minerals with inclusions of other minerals; 3) stony-irons: mixtures.

Practice Test

Multiple choice

1. b	8. a	15. b	22. d	29. b	36. d
2. e	9. a	16. d	23. c	30. c	37. b
3. d	10. b	17. e	24. e	31. e	
4. a	11. b	18. d	25. e	32. d	
5. d	12. c	19. b	26. e	33. a	
6. e	13. d	20. c	27. b	34. a	
7. c	14. c	21. d	28. b	35. c	

True/false

1. F (elliptical)	6. T	11. F (highlands)	16. T
2. F (Saturn)	7. F (low)	12. T	17. T
3. T	8. T	13. F (nebular)	18. T
4. F (smaller)	9. T	14. F (Jupiter)	19. T
5. F (Mercury)	10. F (one)	15. T	20. T

Written questions

1. Compared to the Jovian planets, the terrestrial planets are small, less massive, more dense, have longer rotational periods and shorter periods of revolution. The terrestrial planets are also warmer and consist of a greater percentage of rocky material and less gases and ices.

2. About 5 billion years ago a huge cloud of gases and minute rocky fragments, called a nebula, began to contract under its own gravitational influence. As it contracted, the rotation caused the cloud to assume a disk-like shape with the protosun located at the center. Within the rotating disk, small eddy-like contractions formed the nuclei from which the planets would eventually develop. As the temperature began to drop, materials in the disk began to condense into small rocky and icy fragments. These fragments in turn were swept up by the protoplanets. Because of the high temperatures in the inner solar system, the innermost planets are made mostly of rocky material and lack the gases and ices which are the main constituents of the outer planets.

3. Meteoroid is the name given to any extraterrestrial particle that enters Earth's atmosphere, generally burning to produce a brilliant streak of light called a meteor. On occasions when meteoroids reach Earth's surface, the remains are termed meteorites.

CHAPTER SIXTEEN

Vocabulary Review

1. magnitude
2. irregular galaxy
3. white dwarf
4. H-R diagram
5. hydrogen burning
6. Degenerate matter
7. Stellar parallax
8. nebula
9. supergiant
10. emission nebula
11. protostar
12. neutron star
13. red giant
14. supernova
15. dark nebula
16. pulsar
17. Hubble's law
18. apparent magnitude
19. elliptical galaxy
20. light-year
21. bright nebula
22. black hole
23. reflection nebula
24. spiral galaxy
25. absolute magnitude
26. interstellar dust
27. giant
28. main-sequence stars
29. planetary nebula
30. Big Bang
31. Local Group
32. Doppler effect
33. galactic cluster

Comprehensive Review

1. The nearest stars have the largest parallax angles, while those of distant stars are too slight to measure.
2. The difference between a star's apparent magnitude and its absolute magnitude is directly related to its distance.
3. 1) how big the star is; 2) how hot the star is; 3) how far away the star is
4. a) less than 3000 K; b) between 5000 and 6000 K; c) above 30,000 K
5. Binary stars orbit each other around a common point called the center of mass. If one star is more massive than the other, the center of mass will be located closer to the more massive star. Thus, by determining the sizes of their orbits, a determination of each star's mass can be made.
6. 1) luminosity (brightness); 2) temperature
7. a) E; b) B; c) D; d) A
8. letter C
9. B through C through D represents the main sequence
10. 1) Emission nebulae absorb ultraviolet radiation emitted by an embedded or nearby hot star and reradiate, or emit, this energy as visible light. 2) Reflection nebulae merely reflect the light of nearby stars.
11. Dark nebulae are produced when a dense cloud of interstellar material is not close enough to a bright star to be illuminated.
12. Hydrogen burning occurs when groups of four hydrogen nuclei are fused together into single helium nuclei in the hot (at least 10 million K) cores of stars.

13. a) white dwarfs; b) giants followed by white dwarfs, often with planetary nebulae; c) supernovae followed by either a neutron star or black hole
14. line will go from dust and gases to protostar, to main sequence star, to giant stage, to variable stage, to planetary nebula stage, to white dwarf stage, to black dwarf stage
15. The large quantities of interstellar matter that lie in our line of sight block a lot of visible light.
16. The Milky Way is a rather large spiral galaxy with at least three distinct spiral arms whose disk is about 100,000 light-years wide and about 10,000 light-years thick at the nucleus.
17. 1) Irregular galaxies, about 10 percent of the known galaxies, lack symmetry. 2) Spiral galaxies are typically disk-shaped with a somewhat greater concentration of stars near their centers. Arms are often seen extending from the central nucleus, giving the galaxy the appearance of a fireworks pinwheel. 3) Elliptical galaxies, the most abundant group, are generally smaller than spiral galaxies, have an ellipsoidal shape that ranges from nearly spherical, and lack spiral arms.
18. spiral galaxy
19. When a source is moving away, its light appears redder than it actually is because the light waves appear lengthened. Approaching objects have their light waves shifted toward the blue (shorter wavelength). Large Doppler shifts indicate higher velocities; smaller shifts indicate slower velocities.
20. Hubble's law states that galaxies are receding from us a speed that is proportional to their distance.
21. According to the Big Bang theory, the entire universe was at one time confined to a dense, hot, supermassive ball. About 20 billion years ago, a cataclysmic explosion marking the inception of the universe occurred. Eventually the material that was hurled in all directions from the big bang cooled and condensed, forming the stellar systems we now observe fleeing from their birthplace.

Practice Test

Multiple choice

1. b	7. c	13. a	19. a	25. e	31. b
2. d	8. c	14. a	20. a	26. e	32. e
3. e	9. a	15. b	21. c	27. d	
4. b	10. d	16. e	22. c	28. d	
5. e	11. b	17. c	23. b	29. c	
6. a	12. b	18. a	24. d	30. b	

True/false

1. F (rapidly)	7. T	13. F (Bright)	19. F (Elliptical)
2. T	8. T	14. F (distance)	20. F (brighter)
3. T	9. T	15. T	21. F (high)
4. F (distance)	10. T	16. T	22. F (binary)
5. F (Apparent)	11. T	17. T	23. T
6. T	12. T	18. F (light-years)	

Written questions

1. Following the protostar phase, the star will become a main-sequence star. After about 10 billion years (90 percent of its life) the star will have depleted most of the hydrogen fuel in its core. The star then becomes a giant. After about a billion years the giant will collapse into an Earth-sized body of great density called a white dwarf. Eventually the white dwarf becomes cooler and dimmer as it continually radiates it remaining thermal energy into space.

2. 1) A low-mass star never becomes a red giant. It will remain a main-sequence star until it consumes its fuel, collapses, and becomes a white dwarf. 2) A medium-mass star becomes a red giant. When its hydrogen and helium fuel is exhausted, the giant collapses and becomes a white dwarf. During the collapse, a medium-mass star may cast off its outer atmosphere and produce a planetary nebula. 3) A massive star terminates in a brilliant explosion called a supernova. The two possible products of a supernova event are a neutron star or a black hole.

3. According to the Big Bang theory, the entire universe was at one time confined to a dense, hot, supermassive ball. About 20 billion years ago, a cataclysmic explosion marking the inception of the universe occurred. Eventually the material that was hurled in all directions from the big bang cooled and condensed, forming the stellar systems we now observe fleeing from their birthplace. One of the main lines of evidence in support of the Big Bang theory is the fact that galaxies are moving away from each other as indicated by the red shifts in their spectrums.

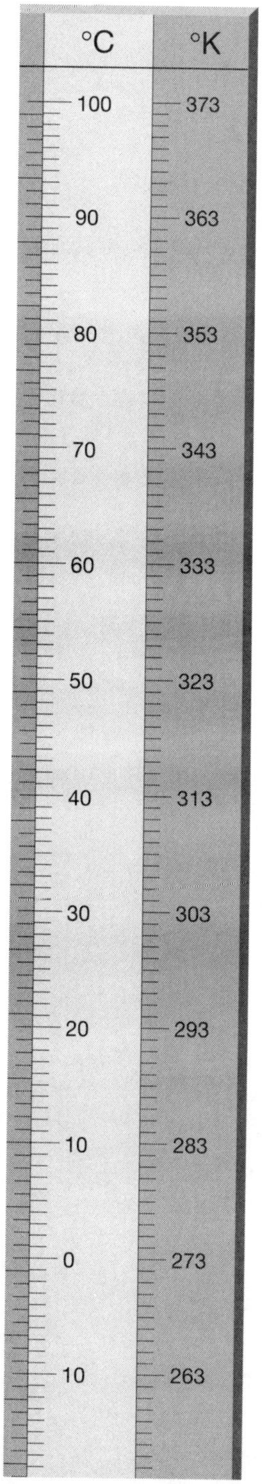
°C
°K
100
373
90
363
80
353
70
343
60
333
50
323
40
313
30
303
20
293
10
283
0
273
10
263